J. Langdon Down, Henry Power

The Skin and its Troubles

.

J. Langdon Down, Henry Power

The Skin and its Troubles

ISBN/EAN: 9783744745918

Printed in Europe, USA, Canada, Australia, Japan

Cover: Foto ©berggeist007 / pixelio.de

More available books at **www.hansebooks.com**

Royal 16mo., cloth.   Price One Shilling each.

HEALTH PRIMERS.

*A SERIES OF SHILLING VOLUMES ON SUBJECTS CON-
NECTED WITH THE PRESERVATION OF HEALTH, WRITTEN
AND EDITED BY EMINENT MEDICAL AUTHORITIES.*

The following Volumes are now ready :—

Premature Death : its Promotion and Prevention.

Alcohol : its Use and Abuse.

Exercise and Training.

The House and its Surroundings.

Personal Appearances in Health and Disease.

Baths and Bathing.

The Skin and its Troubles.

The Heart and its Functions.

To be followed by—

The Nerves.

The Ear and Hearing.

The Head.

Clothing and Dress.

Water.

Fatigue and Pain.

The Throat and Voice.

Temperature in Health and Disease.

Health of Travellers.

Health in Schools.

The Eye and Vision.

Breath Organs.

Foods and Feeding.

And other Volumes.

London :

DAVID BOGUE,

3, St. Martin's Place, Trafalgar Square.

MR. DAVID BOGUE begs to announce that the business, which since the decease of Mr. Hardwicke he has carried on at 192, Piccadilly, under the style of HARDWICKE & BOGUE, is now removed to more commodious premises at No. 3, ST. MARTIN'S PLACE, TRAFALGAR SQUARE, W.C., and will after this date be continued in the name of MR. DAVID BOGUE only. ·

July 1*st*, 1879.

Health Primers.

Health Primers.

EDITORS.

J. LANGDON DOWN, M.D., F.R.C.P. | HENRY POWER, M.B., F.R.C.S.

J. MORTIMER-GRANVILLE, M.D. | JOHN TWEEDY, F.R.C.S.

THE SKIN
AND ITS TROUBLES.

CONTRIBUTORS TO THE SERIES:

W. H. ALLCHIN, M.B. Lond., M.R.C.S.

G. W. BALFOUR, M.D. St. And., F.R.C.P. Edin.

J. CRICHTON-BROWNE, M.D. Edin., LL.D., F.R.S. Edin.

SIDNEY COUPLAND, M.D. Lond., M.R.C.P.

JOHN CURNOW, M.D.Lond., F.R.C.P.

J. LANGDON DOWN, M.D. Lond., F.R.C.P.

R. FARQUHARSON, M.D. Edin., M.R.C.P.

TILBURY FOX, M.D. Lond., F.R.C.P.

J. MORTIMER-GRANVILLE, M.D. St. And., F.G.S., F.S.S.

W. S. GREENFIELD, M.D. Lond., F.R.C.P.

C. W. HEATON, F.C.S., F.I.C.

HARRY LEACH, M.R.C.P.

G. V. POORE, M.D. Lond., F.R.C.P.

HENRY POWER, M.B.Lond., F.R.C.S.

W. L. PURVES, M.D. Edin., M.R.C.S.

J. NETTEN RADCLIFFE, Ex-Pres. Epidl. Soc., &c.

C. H. RALFE, M.A., M.D. Cantab., F.R.C.P.

S. RINGER, M.D. Lond., F.R.C.P.

JOHN TWEEDY, F.R.C.S.

JOHN WILLIAMS, M.D. Lond., F.R.C.P.

London:
DAVID BOGUE,
3, ST. MARTIN'S PLACE, TRAFALGAR SQUARE, W.C.
1879.

LONDON:
PRINTED BY WILLIAM CLOWES AND SONS, STAMFORD STREET
AND CHARING CROSS.

CONTENTS.

THE SKIN AND ITS TROUBLES.

CHAPTER I.

INTRODUCTORY REMARKS.

THE knowledge possessed by the public of the important part played by the skin in the human economy has been, and is still, to a great extent of the most limited description. But thanks to the happy spread of education throughout the country by books, parliamentary measures, and the energy of philanthropists, those conditions of health, which lie in our own hands, are becoming more widely known. Not only is the skin a very important organ as regards its functions, but it excites peculiar interest from its position at the exterior of the body, in that its condition often serves as an index of the departure from a healthy standard of any internal organ, and is also the channel whereby so many powerful influences from without act on the body, promoting health or exciting disease.

In ancient nations, such as the Egyptians, Hindoos, and Hebrews, personal cleanliness and other sanitary

principles were exercised under the form of religious
ceremonies, and the priest was the physician.   Under
the Mosaic economy, ceremonial purification was en-
joined on the Jews, and many other nations possess,
like them, most minute regulations respecting ablution.
In some nations, as the Roman, where the bath acquired such a preponderating position in social life,
cleanliness was the outcome of the laws of society which
made the bath the fashionable and pleasurable resort.
In later years amongst the more civilized nations also
personal cleanliness has been cultivated more in defer-
ence to society, which demands a clean inodorous skin
and prepossessing exterior, than the duty of obeying
one of the first principles of health.   If we have gained
in precise knowledge of the importance of the skin, we
have as certainly fallen back in the care and trouble we
bestow on it.   The time is, however, beginning to dawn
when these matters shall be carried out on the true founda-
tions of physiological knowledge, which is steadily fashion-
ing the habits of Englishmen in the cultivation of health.

The practical outcome of an increasing knowledge has
been seen in the Public Baths and Wash-house Move-
ment, which has had such an enormous influence on the
habits of the people since the first bath of the kind in
London was established in 1845 in the Docks.   It has
been remarked that "it is too little to say that these

institutions are calculated to carry cleanliness into the humble abode of the labouring classes; they do infinitely more than this; they pour forth a stream of health, of happiness, of thoughtfulness; they elevate the moral position of those by whom they are used, and the number-less benefits which they confer are not only enjoyed by those who receive, but are reflected upwards and around upon society at large." If the non-medical reader wants to find facts detailed as to the well-known degrading influences of want of personal cleanliness on the moral nature, and the intimate connexion existing between a clean skin, a clean home, and morality, let him turn to Mr. Chadwick's Report on "The Sanitary Condition of the Labouring Population of Great Britain, in 1842." There he will also gather how much still remains to be done, notwithstanding the great advances made since that date, in bringing the *means of cleanliness* to every home, by developing the erection of baths of all kinds and wash-houses more completely, by bringing water in abundance to every door, and by seeing that every man, whatever his condition in life, has a home that can be kept clean if he will. For although a general steady improvement has been effected amongst all classes of the population in the filthy habits and dwellings of the middle ages, the more rapid advance of the last thirty years has not affected the very poor equally with the middle classes. Mr. Chadwick says :—" No previous investigations had led me to conceive the great extent to which the labouring classes are subjected to privations, not only of water for the purpose of ablution, house-cleansing, and sewerage, but of wholesome water for drinking, and culinary purposes." And a well-known medical officer in the east of London reported that, owing to the scarcity of water, the filth of their dwellings was excessive, and so great was their personal uncleanliness that he could tell at the distance of a flight of steps if any one of the poor was in his surgery. No one at the present time can visit the out-patient rooms of our great hospitals without being struck by the filthy habits, as regards cleanliness, of the poor of this metropolis, and the potent influence such habits exert in the causation of disease. Artisans live on, especially in manufacturing districts, begrimed with the accumulated dirt of years, and never think of cleansing themselves, with the exception perhaps of their face. Employers of labour have the power in their hands of greatly influencing the cleanly habits of the workpeople, by direct encouragement and the supply of the means of cleanliness. Many workmen in special trades break down in health from the simple neglect of the practice of washing their hands before meals, whereby they take into their system lead, mercury, arsenic, and other poisons used in their work.

For the benefit of those who have the means of indulging in a bath each morning, and do not do so, we will here quote an eloquent passage from the writings of a distinguished philosopher, Mr. Bain. He says :—

"Next to eating and sleeping, the bath may be ranked among the very foremost of the necessaries and supports of life. It is of far higher consequence, and of more general utility than any kind of manual exercise, gymnastic, or sport. It affects the system more powerfully than these, even in the very points wherein their excellence consists; and it is applicable in a thousand circumstances where they are not. It does not supersede, but it ought to come before, these other practices. A place should be therefore found for the bath among the regular occupations of life; it ought to be a permanent institution, ranking immediately after the prime necessaries of our being. Either daily, or several times a week, should every one repair to it, in some shape or other, either at morn, mid-day, or evening, according to strength and leisure. There certainly does not exist a greater device in the art of living, or a greater instrument for securing a vigorous and buoyant existence. It is one of the most powerful diversions to the current of business occupati n; it can suspend for a time the pressure of our pursuits and anxieties, and return us fresh for the enjoyment of our other delights. To the three varieties of state which our bodies pass daily through—eating, working, sleeping—it would add a fourth, luxurious in itself, and increasing the relish for all the rest. It would contribute to realize the perfect definition of a good animal existence, which is, *to have the appetite always fresh for whatever is before us.* The health of the mind must be based in the first place on the health of the body; mental occupation and refined enjoyments turn into gall and bitterness if they are not supported by the freshness and vigour of the physical frame."

Amongst other important influences operating on the skin in a baneful manner, unless properly regulated, are the extremes of temperature to which we are subjected by the character of the climate we live in, by the changes in the weather from day to day and hour to hour, and the constant variations, often extreme in degree, to which we make ourselves liable by our habits. If want of cleanliness is a frequent source of disease, certainly exposure, especially to cold and damp, is, perhaps, a still more prolific cause. Diseases of the kidneys and lungs, which are so exceedingly common in this country, stand prominently out as due to, or aggravated by, this cause; and it is the object of clothing, the principles of which we shall lay down, to protect us from the effects of these constant and sudden variations of temperature. Unfortunate beings in thousands, we know, are incapable through poverty of providing themselves with proper clothing; but even those who are better circumstanced often fall victims to cold and damp; and in the selection and wearing of their clothes shew themselves quite ignorant of such rudimentary physiological knowledge as every educated person ought to possess.

We shall devote a special section to some preventable diseases of a special nature, such as those caused by the wearing of certain articles of dress coloured with noxious drugs, and the handling of poisonous articles in various manufactures. We shall also point out the ill-effects of various common customs, such as the use of hair-dyes and cosmetics.

It is necessary to firmly seize the idea that the skin is an organ of such importance that, unless we maintain it by proper habits in due activity, and also protect it from injurious external conditions, it becomes either itself the seat of a number of *preventable* diseases tending to embitter or shorten life, or the inlet whereby they gain access to internal organs. It has often been pointed out that whereas man thoroughly appreciates the importance of the skin in his horse, and exercises the greatest possible care and expends much time and money in protecting the animal's skin from chill and in assiduous cleansing, yet in his own offspring, and perhaps his own person, he systematically neglects these principles.

Lastly. In this little brochure we shall call especial

attention to the subject of the Hair; the conditions of its health, and the means of cure of its disorders. We have particularly insisted on the close connexion existing between hair defects and general ill-health of various kinds, and the injurious results that follow from many of the popular practices in vogue for remedying them. The truth is, that physicians have too long treated this subject with contempt, and the public, in consequence, turns for relief to ignorant people. A very distinguished physician, Dr. Graves, said : " To many it may appear trifling and beneath the dignity of a practical physician to dwell so much on this topic ; but, in truth, mankind have always attached much importance to this ornament of the human body ; and grey hairs and baldness are to many quite as appalling as real disease, and even death."

CHAPTER II.

THE STRUCTURE OF THE SKIN.

IN order to understand how the skin can be kept in good condition and working order, and also be guarded against injurious influences acting either from without the body or within the system itself, it is necessary to be acquainted with its mechanism and functions. We shall, therefore, briefly point out the structure of the skin and the work allotted to it to perform in the human economy. The skin forms a thin, but tough and elastic, investment to the whole external surface of the body. It binds together and protects the delicate structures which lie beneath, completing the beautiful natural configuration of the various regions, and imparting a smooth, soft, external finish to the body. The skin is variously marked with lines in the flexures of the joints and palms of the hands from the incessant folding of the skin, and it is thicker and stronger where increased protection is necessary, as on the palms and soles, and in the nails, which latter parts are to be regarded as modifications of the upper skin or cuticle. In young children and women the skin is softer, smoother, and more delicate than in males; in the middle period of life it is coarser, whilst in old age it gets wrinkled and loses tone from the disappearance of the underlying padding of fat and the diminution in the proper nutrition of the parts. The skin is everywhere firmly connected with the underlying flesh, yet loosely and delicately, as seen on pinching it up, so as to permit of its easy and complete adaptation to the incessant and numberless movements of the various parts and muscles which it covers.

When sufficiently magnified by the microscope, we see that this membranous covering called the skin consists, like the rest of the body, of two sorts of elements—a multitude of little bodies called *cells*, and interlacing bundles of fine threads or *fibres*. Thus built up, the skin is said to be formed by two main layers most intimately joined together; firstly, a superficial dryer and denser one called the *epidermis, scarf-skin*, or *cuticle*, which is the thin skin we see removed by slight abrasions, or raised by the action of a blister in a living body; secondly, a moister, softer, and thicker underlying one known as the *true skin, derma*, or *corium*.

The *scarf-skin* is formed exclusively of more or less rounded cells, packed side by side, and getting flattened out into little scales as they reach the surface. As will be explained, there are no blood-vessels or nerves in this layer, so that you may cut it or pinch it without producing pain or bleeding. The scarf-skin forms a further protection to the more delicate true skin, with its vessels and nerves, and by interposition prevents the too rapid evaporation of moisture, and at the same time the indiscriminate imbibition by the vascular cutis. It also acts as a padding or damper between the influences of the outside world and the nerves. Immediate contact with the nerves causes pain, as we are made aware of in abrasions, but when the insensible cuticle is interposed the delicate sensations of touch are appreciated. The scarf-skin becomes thickened, for the increased protection of the underlying parts, by the prolonged stimulation of local friction and pressure, as seen in labourers, boatmen, sailors, needlewomen, and so on. The action must be gradual to produce this effect, for too free and unusual stimulation only causes sharp irritation and the outpouring of fluid between the two layers of the skin to form a blister. As the scarf-skin has no blood-vessels of its own to feed it, its most superficial flattened cells soon wither and, as they get farther and farther away from the sources of nourishment, they are being continuously shed as fine scurf, and are as regularly recruited by successors from below. The deeper part of the scarf-skin, where it becomes continuous with the true skin, is called the *rete mucosum* or *layer of Malpighi*, for here the cells have a somewhat different arrangement. This is the part of the skin where the pigment accumulates in the cells, which gives the black, copper, and other colours to different races, according to its abundance ; and it is at this point, also, that the workshops exist for the manufacture of new cells, to be successively pushed up to the surface of the scarf-skin and to take the place of those shed.

The *true skin* is fibro-cellular in structure, i.e., it is composed of interlacing bundles of fine threads, and contains, packed in its meshes, cells, nerves, blood and lymph-vessels, and glands, which we shall describe directly. Where it joins the pigment-layer of the scarf-skin, it is raised up into little finger-like processes, like

miniature mountains, called *papillæ*, in which the nerves
of touch end.   These projections raise up the scarf-skin
and give their form to it, as may be readily observed by
examining the ridges on the palm of the hand.   Deeper
down, at the bottom of the true skin, we meet with collections of oil-containing cells, which in the mass constitute
what we know as *fat*, and serve, amongst other purposes,
as stored-up nutriment and a protective padding, both
against external hurt and cold.

There also exist, studding over the whole surface of
the body, little pouches of the skin lined by very delicate
involutions of the scarf-skin, which project downwards
into the true skin, just like the finger of a glove turned
inside out.   These are known popularly as " the pores of
the skin," otherwise as *follicles* or *glands.*   The little
tubes or glands are not, as was supposed, complete per-
forations of the skin, but are closed or blind at their
deeper or inner ends, and are all constructed on one and
the same principle, though slight differences in detail and
use divide them distinctly into two groups.   Thus, in one
class they are about a quarter of an inch long, and terminate by having a considerable length of their tube
coiled into a bundle ; these filter off, from the blood, the
perspiration or sweat, and are called *sweat-glands* (see
Fig. 1).   In the other class these tubes do not reach so
deeply into the true skin, and their blind ends are pouched

out into a lot of little bags in a manner that at once reminds one of a cluster of grapes; their office is to extract from the blood an oily matter, and they are called oil or *sebaceous glands.* This oily matter lubricates the superficial cells of the scarf-skin, and so prevents their too rapid drying up and shedding, besides keeping the surface generally soft and supple. No doubt, also, it somewhat impedes the evaporation of water and the absorption of substances.

There are also implanted all over the surface of the skin, though differing in quantity, quality, and colour in different parts, people and animals, *hairs.* These, when carefully examined and watched during their growth, are found to be nothing else than altered scarf-skin, the cells of which are peculiarly packed together and flattened out into

Fig 1.   C

fibres. They grow from the bottom of little tube-like depressions of the skin, similar to those already spoken about. The structure and use of these hairs and the nails will, however, form the subject of some special remarks in a separate chapter.

As regards the number of "pores" or glands of the skin, and the important part they play in the economy of the body, it may be mentioned that a magnifying glass will disclose on the ridges on the palm of the hand over 3000 openings of the sweat tubes in every square inch. There are also great numbers on the soles of the feet; and even where least numerous, as on the back, there are about 400 present to every square inch of surface; indeed, it has been calculated that there are 2800 sweat glands on the average in every square inch of the body, making a total in an average-sized man of about 7,000,000, which is equivalent to twenty-eight miles of the tubing, since each tube is a quarter of an inch long. It is evident that an enormous amount of perspiration may be filtered off from the blood by this extensive surface. The sebaceous or oil glands, with few exceptions in certain parts of the body, always open into the mouth of the hair-tube, so that a notion may be easily formed of their numbers also. They do not exist on the palms of the hands and soles of the feet, where the sweat glands are so numerous.

Two other very important elements of the structure of

the skin remain to be noticed, viz., the vessels and nerves of the skin.

The *vessels* are of two kinds, those circulating the *blood*, and those containing a fluid, derived principally from the blood, called *lymph*.

The *blood-vessels* are elastic tubes with little muscles in their walls, and proceed from the heart, which is the pumping engine to convey nutriment in the blood to every part of the body. The vessels divide into branches, after the manner of a tree, as they course along; and when those destined for the skin reach that organ they divide into innumerable little twigs called "capillaries," only visible with the microscope. *These never reach the scarf-skin*, but run along and form a very close network in the true skin, to distribute nourishment thoroughly and uniformly everywhere, and with especial liberality to such important workshop parts as the papillæ and roots of the hairs, and to the oil and sweat-forming glands. This done, the little vessels commence to join together again, forming larger and larger tubes, which convey the now impoverished blood back again to the heart and lungs to be purified and replenished for re-circulation. So numerous are the blood-vessels in the skin, that it is impossible "to direct the point of the finest needle into any spot without puncturing a vessel and drawing blood." It is to the red colour of the blood in the true skin, toned down by the interposition of the cuticle, that the beautiful pink flesh tint is due; and the more blood present in the skin the redder is the hue of the surface, for the blood-vessels are capable of expanding and contracting, and so altering the quantity of blood circulating in the skin.

The lymph vessels or *lymphatics* of the skin may be displayed in endless profusion by injecting them with quicksilver. They do not commence quite as tubes, but chiefly as little spaces between the cells and fibres of the interlacing network and gradually form tiny tubes, as the blood-vessels do, to carry away the lymph into the interior, where it is poured into the blood together with the nutrient fluids absorbed from the digestive tract.

These two kinds of vessels act after the fashion of irrigation and drainage works. The fluid part of the blood, as it circulates, soaks through the walls of the capillaries and bathes every cell and fibre, just as the intervening land between the streams is soaked in a water-meadow. A great part of this nourishment is taken up by the elements of the skin and used for producing new cells, and in the wear and tear of the work of the body, but the used-up or effete material is exchanged back into the blood to be got rid of. The excess not made use of is drained off principally by the lymphatics to be worked up and used again, whilst the waste matters are filtered off, or got rid of, or excreted, as it is called, by the little glands of the skin and by other internal organs, such as the liver, kidneys, bowels, &c.

Lastly come the all-important *nerves*. They are cords proceeding from the spinal marrow and brain, and, getting finer and finer by the offset of branches, they end in little twigs in the various parts of the body in different proportions. The skin is profusely supplied with them, and some parts in a remarkable degree, such as the palms of the hands, for the special elaboration of the sensation of touch.

These particular nerves of touch terminate in the papillæ, and their ends are protected and wrapped up in a very peculiar manner to form a knot-like body. The uses and purposes of the nerves are very multifarious, and indeed they preside over and direct every process that goes on in the skin. They may be likened, in a rough manner, to telegraph wires radiating to all parts of the country from a central office. Each "wire" consists of a number of smaller ones, with different functions, bound side by side together into a bundle. Thus some nerves convey up from the skin to the central office the sensation of touch, or heat, or cold, and then an influence is at once sent down by other nerves to the muscles to contract and move the part away, or down some of the other sets of nerves for special purposes. Some nerve fibres, called *trophic*, have a special duty to preside over the selection of nourishment and growth of the tissues.

Others, called *secretory*, are in intimate relation with the filtering of the sweat and oil from the blood by the glands. Whilst others again, called *vaso-motor*, are distributed to the little muscles encircling the blood-tubes, and have the special duty of governing the calibre of the vessels and causing the latter to become larger or smaller according to circumstances : so these regulate the quantity of red blood admitted to the skin. A familiar instance of the influence of the nerves on the blood-vessels is seen in the production of blushing or pallor under nervous excitement, or changes of temperature ; on the secretory glands in the profuse perspiration of fright. It is important to remember these different functions of the nerves, as in diseases of the skin we incessantly come across illustrations of their actions.

CHAPTER III.

THE FUNCTIONS OF THE SKIN.

THE skin may be said to perform six chief functions, some of which have already been incidentally alluded to, and upon their proper fulfilment depend in great measure some of the most important conditions of health of the body.

Firstly, the skin forms a *protective and supporting covering* to the whole external surface of the body.

Secondly, it is the seat of the termination of certain nerves in peculiar abundance, whereby all the degrees and varieties of the *sensations of touch and temperature* are appreciated. In the skin these nerves of sensation are distributed and protected, and the accurate impressions made on them by external objects are at once transmitted to the central nervous organs to be acted on if need be. The properties of these nerves come into play in our every action unceasingly; indeed their due exercise is absolutely essential to our continued existence. Striking examples of the necessity of their action are afforded by severe frostbite or burning occurring in a palsied limb without the knowledge of the individual.

Thirdly, the skin possesses the *function of absorbing matters* brought into contact with it, *i.e.*, of sucking them, as it were, into the lymph and blood vessels. This, although of secondary importance, is a fact which should not be lost sight of, as it has practical applications. It is well known that water is taken in through the skin during immersion in a bath, and also that various liquid and solid drugs, such as Iodine and Mercury, thus enter the system when rubbed in contact with the skin; and vapours also are absorbed, as in mercurial fumigation. There is no doubt, therefore, that the skin does absorb matters, but the extent to which it occurs varies very greatly, according to the nature of the substance and the state of the skin. The vascular true skin however absorbs very rapidly and indiscriminately, as is a matter of common experience in the poisoned wound following an abrasion. It is only the coating of cuticle which stops this, but probably not entirely for certain substances. The moral to be learnt is to be careful what we place in contact with our skins, and to remember that it is possible that this is one of the inlets through which poisonous malarious gases and the products of filth and overcrowding make their insidious entrance.

Fourthly, the skin acts as a *Breathing Organ* accessory, and playing a secondary part, to the lungs. The latter, in breathing, take in a large quantity of a gas called Oxygen from air, and in exchange receive from the blood and exhale the impurity called Carbonic Acid, mixed with the vapour of Water and some Organic Matter. So the skin in like manner takes in a little Oxygen from the air and gives out a little Carbonic Acid and a great deal of vapour of Water.

Fifthly, the skin is the *great regulator of the temperature of the body*, and in this way. The oxygen mentioned above as taken into the body when we breathe, burns up by chemical action the food taken into our blood by way of the digestive tract, and in the process produces a great deal of heat, warming the blood, which then rapidly conveys this heat to every part of the body. Now it is necessary for the healthy life of the tissues that they should be always kept by the blood at a certain heat known as blood-heat. And as a fact the blood marks, whether a person be in the Tropics or the Arctic regions, under the most variable conditions of climate, always about 99 or 100 degrees on the Fahrenheit thermometer. If there were no escape for this constant production of heat, the body would get very hot indeed, but channels of loss are provided in several directions. Everything hot that passes from the body, and every contact with a colder external body conveys away a certain amount of heat, besides the general cooling of the body by radiation, as a hot cannon-ball cools in air. But it is left to the skin to regulate the body at a constant temperature with nicety, and it does this in two chief ways. First, the fluid from the blood that bathes the skin is constantly passing off in the form of vapour as from a wet cloth suspended in the air, and this evaporation, as it is called, cools the body; for water can be even frozen by the cold produced by its own evaporation. Second, a great amount of heat is lost in like manner by the evaporation of the sweat secreted by the sweat glands. When the air is cold, the nerves apprise the central nervous system of the fact, and as a consequence the blood-vessels are made to contract, so that less blood enters the skin, and consequently there is less water to evaporate and less heat lost. Conversely, when the air is warm, the vessels hold more blood in the skin, and greater cooling goes on. But moisture in the air also plays a very important part, for the greater its amount present the less free is the evaporation and cooling, so that in a very hot moist room the vapour has difficulty in escaping from the body, and forms beads of perspiration on the skin. The quantity of sweat must not be judged of by what is seen, or the *sensible* perspiration, for by far the greater part escapes as unseen vapour, or *insensible* perspiration.

Sixthly and lastly, the skin is one of the three great organs of the body concerned in the *purification of the blood* by the getting rid, or elimination, of the used-up or waste materials, the accumulation of which in the system is followed by injurious consequences. It is by means of the sweat principally that this elimination by the skin is effected. This fluid, when filtered off, as it were, from the blood, carries with it the waste products in solution, just as chalk may be held in hard water. The chief of these excreta formed in the body are Carbonic Acid, Water and some other materials, the most important of which is Urea. The skin in health gets rid of a great deal of Water, a little Carbonic Acid, and a small quantity of Salts : but in disease it tries to excrete various abnormal and injurious products.

We will here take occasion to make a few remarks on *the compensatory or reciprocal action and sympathy existing between the skin and other organs.* If by any cause, such as a chill, the elimination of waste products by the skin be checked, they must either accumulate in the blood or be got rid of by other channels. As a matter of fact other organs in the body, with a community of function, act vicariously and try to do the extra work thrown upon them. These organs are the lungs, intestinal tract, and especially the kidneys. A familiar illustration is the increased action of the latter organs when the evaporation of water from the skin is checked by cold, as on passing from a hot room into the cold air. But although a certain amount of extra eliminatory work can be vicariously performed the quantity is limited, and the effort may damage an over-worked organ, especially when it is already enfeebled or predisposed to disease, causing very serious mischief. If the waste products do accumulate in the blood, they act as a severe source of irritation, and throw out of gear all the healthy processes of the body ; indeed their continued accumulation is incompatible with life. Mischief, more-over is not confined to this head, for besides the non-elimination of waste products there is a state of conges-tion of internal organs induced by the expulsion from the skin by a chill of the great quantity of blood contained in it ; a condition which may end in inflammation. And

again, it is not a mere question of an extra quantity of
blood being forced internally, but there is such a wonderfully close sympathy existing between the different organs
through the agency of the nervous system, that any shock
or bad influence communicated to one part acts preju-
dicially on all. This is well shown in the case of the
stomach and skin. This compensatory action between
the organs may also be illustrated conversely. Thus if a
quantity of fluid has passed from the body in an attack
of diarrhœa, the action of the kidneys is slack; or, as
very commonly witnessed, if the kidneys cannot get rid
of the waste water, and the skin does not compensate
sufficiently, the water collects in the tissues of the body
as dropsy. So when the kidneys are diseased, careful
analysis has proved the existence of the waste product
called Urea coming from the body by the skin in an
abnormal manner.

Certain other organs also, though not employed in
separating from the blood waste or used-up materials,
have certain functions to perform in very intimate connexion with the blood, so that if their action is deranged
impurities at once accumulate; not ordinary waste matters,
but really foreign products. For instance, the liver is
one of the most important organs in the body, and has
a great many complicated functions to perform in working up the nutriment in the blood to a proper condition

for circulation to the tissues. If the liver be disordered substances immediately collect and are circulated with the blood that ought not to be there, causing such diseases as Jaundice, Diabetes, and Gout, and having the most injurious influences on the functions and nutrition of the skin, exciting and intensifying different kinds of inflammation, unbearable itching, &c.

The matter has been summed up in these words: " Thus the skin, the bowels, the lungs, the liver, and the kidneys, sympathise readily, because they have all the common office of throwing waste matter out of the system, each in a way peculiar to its own structure ; so that if the exhalation from the skin, for example, be stopped by long exposure to cold, the large quantity of waste matter which it was charged to excrete, and which in itself is hurtful to the system, will most probably be thrown upon one or other of the above-named organs, whose function will consequently become excited ; and if any of them, from constitutional or accidental causes, be already weaker than the rest, as often happens, its health will naturally be the first to suffer. In this way the bowels become irritated in one individual, and occasion bowel complaints ; while in another it is the lungs which become affected, giving rise to catarrh, or common cold, or perhaps even to inflammation. When, on the other hand, all these organs are in a state of vigorous health, a temporary increase of function takes place in them, and relieves the system, without leading to any local disorder; and the skin itself speedily resumes its activity, and restores the balance among them."

CHAPTER IV.

PRACTICAL APPLICATIONS TO THE CONDITIONS OF DAILY LIFE.

WE now propose to consider the modes in which the considerations set forth in the preceding chapters are to be utilized for the preservation of health, and to point out how the functions of the skin are to be kept in healthy activity so that the body generally may profit.

Cleanliness.—It is simply impossible to overrate the importance of cleanliness, as anyone may observe by noting the number of diseases caused or aggravated by dirt.

The most fitting time for cleansing the skin is on rising in the morning; but before speaking in detail on this point, it will be well to briefly consider the objects of washing. The reader will remember that in describing the scarf-skin we mentioned that it was constantly being shed in the form of fine scurf. Now much of the cuticle is constantly falling off, but a good deal remains entangled in, and stopping up, the mouths of the "pores" or follicles, whilst other portions are as it were stuck on the surface by the lubricating oil, and this part of it gets mixed up with little particles of " dirt " alighting upon the skin from the air. The oil or sebum also of the surface, if left long unremoved and exposed to the air, is liable, especially in some situations, to become rancid, and then is unpleasant to the senses as well as irritating to the nerves of the part. The same may be said in regard to some of the materials deposited or left behind upon the skin by the evaporation of the sweat, so that a good deal of unpleasant matter therefore readily accumulates on the surface if not prevented. When, moreover, we remember that the " dirt " of the air is found by the microscope and other appliances to consist of minute particles of almost a very multitude of foreign matters, from the dust of the roads and " blacks " of smoke to decomposing organic matter and the seeds of ringworm, the objects of cleansing the surface of the skin and opening the pores must be evident. And it is not any special part of the skin only that needs ablutionary measures. These are requisite not only in the case of the more exposed portions such as the face, neck and hands, but to every part, and especially to the regions covered by flannel. Further, certain occupations predispose to unclean skins. To those whose work is amongst metals such as lead, or amongst the dust of coal-pits or warehouses, or in crowded populous foul courts, the necessity is still more urgent. Disease is often directly traceable to this neglect, and the absence of cleanliness in the rearing of children is one of the most fertile sources of disease of the skin. Soap and rubbing are the great allies of water in this matter of the purification of the skin, and for very simple reasons. Soap is composed of a certain amount of alkali—potash or soda—with a less proportion of oil, and the alkali acts by partly softening up or loosening the cells of the scarf-skin and so helping their removal, and partly by getting rid of the surface oil by forming an emulsion with it. The rubbing necessarily removes the softened-up dirt, and stimulates the parts to renewed action. No one, whatever his position in life, need go without the use of a piece of soap and some water, and so give the skin a chance of breathing and getting rid of impurities; the appliances are simple, and the object of the first importance to the health of the body. Cleanliness of the person almost necessarily involves a cleanly home, and to the spread of these principles is due doubtless the extinction in England of some diseases, and the diminished prevalence of others.

The number of ordinary soaps offered to the public is great, and most of them are good. They are all vaunted with varying degrees of success. The great object in selecting a soap is to obtain one that has a moderate quantity of alkali. If the latter is much in excess, the result of its use is that the cuticle is too freely softened away, and the skin is left dry and harsh. Soap with much alkali may, especially in the case of tender and sensitive skins, induce decided inflammation. It is very important that the soap used for children, especially to the face, and in cold weather, should be a mild one, i.e., containing little alkali. Another important point is to take care that the parts to which soap is used are thoroughly dried after having been washed. If with the mildest soap the skin gets dry and chafes, a little greasy application may be used, such as cold-cream or a preparation known as Vaseline. Medicinal soaps are also numerous, but they are only necessary in deviations from health.

Bathing.—It is necessary here to say a word on bathing, including the morning cold bath, hot baths, river and salt-water bathing, and the Turkish bath.

The " morning tub," or sponge bath, is one of the most healthful customs the ordinary vigorous Englishman can delight in, or even the young can be indulged with. It is not only an excellent means of cleansing the skin, but at the same time ought to be so used as to be made to serve as a good nerve tonic and "hardener." The first effect of cold water upon the skin is to stimulate and contract the vessels, but this is soon followed by an increased flow of blood through the skin, accompanied by a sense of warmth and comfort, to which the term "reaction" is

D

applied. Some constitutions are not well affected by a cold bath. Its use is not succeeded by a greater vigour of circulation, i.e., the reaction does not take place. The test of its good or bad influence is to be found in the occurrence or non-occurrence of "reaction," which is indicated by the glow of warmth of the surface, and the vigour and exhilaration induced. If this reaction does not occur friction may induce it, and then the bath may be carefully taken ; but if depression follows with shiverings and persistent cold in the extremities, which are not speedily removed by friction, a tepid bath is advisable in place of the cold bath. In very cold weather too it is best to take the chill off the bath. It is a capital practice to train children in the regular use of the bath fairly or nearly cold ; but great care must be observed in the matter, and the effects carefully watched for. Children are very sensitive and of tender strength; they do not possess the power of reaction that belongs to the adult. All attempts to "harden" them by exaggerated means are most baneful and should be abandoned. The same remarks will apply also more or less to girls, but the adoption by them of a daily cold or tepid bath is most important, and there is great neglect in this matter. Where the simple cold bath cannot be borne, it is often an admirable plan to modify its character by the addition of some saline substance, which then stimulates the skin and enables the bath to be borne. Such things as prepared sea-salt may be employed, and in cases where a tepid bath only can be taken this preparation is a most useful adjunct, because it imparts to the bath a tonic quality which counterbalances the somewhat relaxing effect of the tepid water.

Hot Water Baths are no doubt more efficacious for cleansing purposes than cold water baths; but they are not to be recommended for constant daily use as they are relaxing, and lack the bracing tonic properties of the "cold tub." When used they should not be taken too frequently, not more than once a week or so.

The Turkish Bath is the best of all for cleansing purposes, for it increases the secretion of perspiration, and this fluid softens up the cuticle as it were, and enables the latter to be readily removed. With the various medicinal uses of this form of bath it is not our purpose to deal here. When taken occasionally as a cleansing operation, it should not be indulged in within two and a half or three hours after a meal, and the bather should lie down quietly, never go into too hot a room, never for more than a few minutes into a room of above 150° F., never stay too long in the bath, and should finish by the free application of soap and water and cold douche or plunge. It is by the neglect of these precautions that many persons feel languor, headache, loss of appetite, and the like, afterwards.

It may be well to offer a few remarks on *fresh-water and salt-water plunge bathing.* Salt-water differs from fresh in containing more *salts* in solution, which have a specially stimulating action on the skin, and through that organ on the system generally. Most Englishmen know the extreme pleasure afforded by the exercise of the art of swimming, the pleasant sensations communicated to the body through the skin, and the bracing subsequent effects, when the bath is suitably taken. Nevertheless, nothing is more common than to see bathing performed under the most injudicious circumstances, and in violation of well-known rules which may be touched upon here. Bathing is to be avoided during a period of two and a half hours after a full meal, because during that time digestion is proceeding and would be interfered with. The bather should have a brisk rub down after bathing, and stimulate the circulation by a little exercise in cold weather. The vigorous may bathe on an empty stomach the first thing in the morning if desired, but the water should not be too cold. The weak should be careful not to bathe when they feel in any way exhausted, and in fact no one should bathe when much exhausted or chilled, as the vigour of the system is rarely resuscitated, since no reaction takes place; but the bodily powers are only further depressed, and then colds, coughs, and other local disturbances are apt to occur from the effects of chill upon the skin. No one should after exercise take a bath if he is still perspiring profusely, nor too long after exercise when the body is beginning to chill, but the bath should be taken whilst the body is still comfortably warm. If the bather remains too long in the water the period of reaction passes by and depression ensues and continues. The great point as before remarked is to regulate bathing by the amount of exhilaration and vigour infused into the body, and the glow of warmth on the surface. When this is produced the bather should be satisfied and leave the bath for a moderate amount of friction before dressing.

Clothing.—The objects of clothing are to protect the skin from external hurt, to bedeck ourselves to please the eye, to fulfil the conditions of decency demanded by civilized society, and lastly, by artificial means, to moderate the action of extremes of heat and cold, and equalize the too rapid changes of external temperature to which we are subject. The importance of this subject may be insisted on by such illustrations as the injurious influence on the nerves of the direct action of the sun's rays seen in the nervous depression and sunstroke induced; the production of catarrhs, bronchitis, inflammation of the lungs, kidney disease, rheumatic fever, and neuralgia by the too sudden checking of perspiration and exposure to cold; the causation of our heat-rashes and the more severe kinds of the tropics; the baleful influence of the siroccos, and the rapid equatorial changes of temperature, &c. " People know the fact, and wonder that it should be so, that cold applied to the skin, or continued exposure on a cold day, often produces a bowel complaint, a severe cold in the chest, or inflammation of some internal organ. But were they taught, as they ought to be, the structure and uses of their own bodies, they would rather wonder that it did not always produce one of these effects."

When woollen fabrics are used as clothing the sensation of warmth produced in the body is not communicated from the wool itself, but the garment acts as a screen, and so hinders the heat of the body from passing away too rapidly into the exterior under the cooling influences of moving air or wind. An illustration of this hindrance to the passage of heat is afforded by the common custom of wrapping ice in flannel, though here the heat tends to pass in to melt the ice. Everyone knows that different materials, e.g., copper, iron, glass, wood and clay, allow heat to pass through them with different degrees of facility. This is called the power of " conduction," and this is the great principle applied in the selection of clothes. Hare's fur, eiderdown, and woollen garments are the warmest because the worst conductors, and cotton warmer than linen, and silk cooler than either. Of these materials, either pure or combined together, most of our clothes are made ; for instance, calico is pure cotton, flannel is wool, and merino composed of cotton and wool. In selecting clothes also, it is necessary to pay attention to their texture, not only because rough materials irritate some sensitive skins very much, but because wools, especially when unwashed, have the property of soaking up and retaining moisture in their interstices, and not allowing it to pass through as readily as linen. But further, the *colour* is of importance, for just as a mirror reflects light so do clothes heat. Black and blue are hot colours, for they absorb much heat and do not reflect it, white and grey are the best protection from the sun's rays, i.e., the coolest because they do not absorb heat but reflect it ; and thus colour, and this is at command by dyeing, and material must be combined in the application of these principles to season and climate. Such clothes as skins, leather and india-rubber, are very hot because no winds can blow through them, and the vapour of the perspiration passing off from the body is shut in and condensed.

The fit of clothes to the skin should be attended to. It is important that they should sit loosely and not compress the skin. Lastly, great care must be paid to a proper warmth of dress in infants. They must neither be too warmly clothed as is so often the case, nor, what is of still greater importance, too lightly dressed, with the view of so-called " hardening " them. Cold is a great source of disease in children, for not only do they lose heat rapidly, but their production of it is limited, and their vigour is not so great as an adult.

Diet.—In common with all other parts of the body, the skin derives its nourishment from the circulating blood, and the blood acquires it from the stomach. If the latter organ does not "digest" or prepare the food properly, from, perhaps, irregularities of time and methods of eating, or if unfit substances or liquids, or even an excess of proper things are taken so as to hamper its action, it is evident that improperly prepared nutriment, or even more actively injurious material, must get into the blood, and so reaching the skin derange the normal growth of the cells by irritating the nerves and blood-vessels, or disturbing the action of the glands. The influence of the stomach on the skin in producing and intensifying eruptions is known and recognised as a fact almost universally. It varies from the production of flushing of the face after food to the causation of "nettle-rash" after eating stale fish. And it must be remembered that, similarly, other organs are thrown out of gear, especially the liver, and improper and morbid materials in consequence of this also accumulate in the blood, and thus act injuriously on the skin; so that not only the *direct*, but the *indirect* results of improper diet are considerable. The stomach moreover is wonderfully supplied with nerves, and oftentimes it happens that any bad influence acting on it is at once "telegraphed" or radiated to the skin with a bad effect.

Physicians are only too well acquainted with the fact that all things that induce *dyspepsia* are liable to be followed by disorders of the skin, such as rose-rash or erythema, nettle-rash, acne, eczema, pruritus or itching which may readily become chronic and troublesome, and so forth. This is particularly exemplified in the instance of persons who possess any hereditary tendency to any of the common forms of skin trouble, especially such as are of rheumatic or gouty nature. In such cases errors of diet or excessive diet which would be harmless in the case of others, is followed by mischief in persons of these particular constitutions, and the food should be, under such circumstances, carefully chosen and of simple nutritive character. Sugar, excess of meat, rich things, condiments and stimulants of all kinds are to be used sparingly ; more so, that is, than under ordinary circumstances.

But apart from the influence of the stomach, it is also very important that food should be taken in proper quantity and character, that the skin may be properly nourished ; for it early shows the effects of malnutrition through weakness of the general health in becoming thin, wrinkled, harsh and dry, in the great predisposition to take on diseased action on any provocation, and in the intensification of any existing symptom.

The occurrence of eczema in infants from defective feeding is an apt illustration on this point. The effects of the prolonged abstinence from fresh vegetables in producing scurvy is well known; not that there is, as some people suppose, any exceptional blood-purifying influence in vegetables, but the varieties of food should be properly mixed, and the diet not consist exclusively of meat or rice, for instance.

In the young the bad effects of an ill-regulated diet, especially in the children of delicate and scrofulous parents, are constantly seen by the physician, and there are one or two points of general application which may with advantage be noticed here. The first is, as regards infants, that it cannot be too strongly insisted upon that the natural food supplied by the mother is infinitely preferable to any artificial substitutes. Children do not thrive so well on the latter, but are more liable to become weakly, and as a consequence the disease eczema is of frequent occurrence amongst them; of this there can be no doubt. The practice of many parents of making their infants' food too sweet is a bad one. Sugar in excess induces acidity, which causes a state of blood that speedily irritates the skin. Many parents, and even doctors, have great faith in wine as a means of giving strength to young children; there cannot be a greater mistake, for the wine is often given at the expense of good solid food, and debility results as a consequence, the wine at the same time inducing much irritability and liver disorder indirectly. Wine should only be given to young children as a medicine. Frequently in the delicate strumous children the evil consequences of this deprivation of proper solid food is followed by skin trouble which is in part due to this cause, as in cases of lupus and scrofulous mischief. The too free use of farinaceous foods, to the exclusion of a proper amount of meat, has a like evil tendency. In the instance of growing boys, and especially girls, at or approaching the age of puberty, the regulation of the diet is of the greatest moment. A due amount of meat must be taken; and pastry, puddings, tarts, sweets, and what is generally known as "trash," are to be used in the greatest moderation. It is often at this time of life that hereditary tendencies to scrofulous disease of the skin, and to eczema and other skin troubles show themselves on the surface, because the subjects of them are under—or imperfectly fed. Nothing is more common than the occurrence of acne at this age from debility, or its aggravation by dyspepsia in part produced by careless dietary. In persons of maturer age, high living and over stimulation induce a number of cutaneous troubles, from the fact that the blood current is charged in excess with impurities and the waste products which in the former case are freely introduced into the blood, or in the latter are retained because the organs whose duty it is to remove them are disordered by the too rich food and the free use of stimulants.

Water is a very important article of diet, and the existence of impurities in it is a fertile source of diseases even in this country, with manifestations in the skin. In foreign countries peculiar boils and ulcers are thus caused, and worms introduced into the skin and system.

Exercise.—Exercise increases the vigour of the heart or pumping-engine, and so the briskness of the circulation of blood through the skin. Consequently all the processes of oil and sweat formation, and the elimination of impurities go on more actively. Exercise also is the means of bringing this increased quantity of blood in contact with pure fresh air, and so fulfilling more actively the function of the skin as a breathing organ, whilst at the same time the nerves are stimulated by the moving air. A sedentary life makes the skin and its functions sluggish, as it does with the lungs, and the kidneys suffer from the extra work thrown continually upon them.

It is important to take exercise regularly and in moderation, at as fine and fresh a part of the day as possible, and not too soon after meals, as with bathing. Immediately after exercise, the injurious effects of too rapidly checking the perspiration by sudden immersion in cold water, or sitting in draughts, are to be avoided. The lighter clothes worn during the exercise should be replaced by fresh and warm ones.

The Action of the Bowels.—The bowels serve to convey away the refuse of the food taken into the stomach, and at the same time act as a blood purifier by separating from the blood some of the waste products, although in small amount compared with the kidneys, skin, and lungs. Consequently if the bowels do not act regularly, there is not only danger of its share of waste products not being eliminated, but of the refuse matter being absorbed in the blood and contaminating it. Practically we know that if the bowels are constipated, skin troubles, such as heating of the face, pimples thereon, itching, &c., are induced, or existing conditions of disorder are intensified. Every physician recognises this. Hence the paramount importance of securing daily relief; an omission, in the case of the majority of young women, which is truly disastrous as regards the skin.

_____

CHAPTER V.

SKIN TROUBLES FROM POISONOUS CLOTHING, INJUDICIOUS USE OF DOMESTIC REMEDIES, ETC.

THERE is a class of cases, demanding notice at our hands, in which severe and sometimes serious inflammation of the skin is produced by the use of certain articles of clothing coloured by arsenical, copper, or organic dyes. This is a distinctly preventable class of diseases, and we will, therefore, in a few words put the reader in possession of the main facts of the matter that he may be on his guard, for prevention is decidedly better than the cure of these affections. Some useful comments may also be appended concerning the employment of what may be termed domestic remedies or medicines (such for example as Arnica), which the public are wont to use on their own judgment.

Poisonous Clothing.—A few years since a great stir was made amongst the public and the profession, not only in this country, but also in France and Germany, by the revelations made about the poisonous action of Aniline dyes on the skin. These dyes are manufactured from certain ingredients of coal tar, and almost every shade of brilliant colour seen in the animal and vegetable world has been successfully imitated and reproduced in them. They are largely used on account of their beauty by the colour-printer and the dyer. There are aniline-reds (known as fuschine, coralline, magenta, or rose-aniline), aniline-yellows, aniline-greens, aniline-blues, aniline-lakes, and many others.

Professor Tardieu was one of the earliest to call attention to the evil action on the skin of these dyes in a paper read before the French Academy in 1869. Special reference was made there to coralline sock-poisoning. Professor Viaud-Grandmarais then recorded a case of poisoning from wearing a red flannel waistcoat dyed with aniline red, and several Englishmen published instances which had come to their knowledge of mischief to the skin resulting from wearing the bright orange, yellow, and red coloured socks, which formed at the time so conspicuous an object in the haberdashers' windows. One case was that of a ballet-girl, who wore tights with one of the legs brilliantly coloured, and the skin of which leg was seriously inflamed in consequence. One of the medical journals (*Medical Times and Gazette*, 1869) at this period stated that they had been informed "not only of ladies whose skins had suffered from tinted flannel waistcoats, but seamen whose backs and arms had been excoriated by wearing 'singlets,' i.e., light woollen tunics similarly dyed ; and also of beautiful pink soaps which had irritated the skin." In like manner certain red chest-protectors (*Medical Times and Gazette*, Feb. 27, 1869), red flannel waistcoats, red shirts (*loc. cit.* April 17, 1869), magenta-coloured glove-linings (*loc. cit.*), and the red lining of hats (*Deutsche Wochenscrift*, July 1, 1869) were also shown to have induced skin inflammations of a severe kind. We can confirm by our experience the truth of these assertions by numerous instances, some of which are on record elsewhere.

Now, it has been found by careful examination that in some instances these poisonous articles have yielded arsenic in some form in addition to the aniline compound itself. Aniline is most readily obtained from an ingredient of coal-tar, called Benzol, by oxidation, and it so happens that this process is usually most conveniently carried out by the use of arsenious acid. Hence two factors in the causation of these ill effects exist—the organic dye and arsenic, for it has been clearly shown by Tardieu, Mackey, and others, that the aniline dye itself without the arsenic is quite potent enough to cause serious mischief.

There can be no doubt that the outcry raised had the effect of stopping in a great measure the employment of these dyes in haberdashery especially at first, but, the colours are effective, and on this and other grounds there is every reason to believe that articles coloured by these dyes are still sold. Only a short time ago we were called to treat a very severe case of sock-poisoning of the feet and lower part of the legs in a well-known member of Parliament, who had obtained the socks from one of the best known West End shops. Thus the subject is still one of lively public interest, and we venture to caution our readers against the purchase of any gay-coloured article of dress unless he or she is assured that the dye does not contain any deleterious ingredient. These aniline dyes act with especial intensity in hot weather when the skin is in a state of perspiration, since the acid sweat tends to dissolve out the dye, and so the latter gets absorbed.

The symptoms produced vary somewhat ; usually they consist in redness and staining of the part, followed by swelling, itching, and smarting, with the formation of little blisters,, or vesicles, which break and give exit to a discharge. The part affected then becomes decidedly painful and is occasionally greatly swollen. There is also a good deal of constitutional disturbance, and in fact the sufferer feels quite ill. The peculiar staining of the skin coinciding with the particular hue and pattern (bars, stripes, &c.) of the coloured article at once suggests the cause of the mischief.

In less severe cases an irritant pimply rash only is produced. This is sometimes observed on the forehead as the consequence of the wearing of a hat lined with stuff dyed with irritant colours. In the case of the hand affected by gloves lined with the same the extent and degree of the inflammation will vary.

In meeting these cases the measures required are both internal and external, the former to prevent the spread and lessen the severity of the inflammation by the administration of salines, diuretics, and the regulation of the diet, &c. ; and the latter to soothe and reduce the severity

E

of the swelling, heat, pain, and discharge, and to hasten the healing. This is of course a purely medical topic, and the affection one which ought not to be treated by the sufferer himself; but we may just say this, that where aniline poisoning occurs, before the arrival of the medical man, it is advisable at the very first occurrence of symptoms of disease to wash the parts well with a little simple soap and water (the latter containing a little carbonate of soda), then to put a good poultice on for a short time, and finally to wrap the part up in an embrocation consisting of equal parts of olive oil and lime water, or to flour the parts after washing them gently.

Arsenical and Copper Poisoning.—It is within the personal knowledge of most medical men that colouring matters containing arsenic or compounds of arsenic and copper, frequently produce irritation of the skin. These arsenical dyes (generally *green*) are found in ball-wreaths, clothes, house-papers, artificial flowers, toys, and confectionery. When brought into contact with the skin, they induce redness and an eruption of pimples, and sometimes blisterings, mattery heads, and severe ulcers, or superficial gangrene. These accidents are mostly seen in those who are occupied in the manufacture of these different articles, but in others also who make use of the latter. The local skin trouble is often attended with serious evidence of disturbance of the general health, such as headache, giddiness, obstinate dry cough, and other " pernicious insidious effects." In cases, therefore, of unaccountable or unusual skin trouble, it is always desirable to remember that the cause may be found in the use of articles of fashion containing arsenical colouring substances. Van der Broeck (Ranking's Abstract, vol. i., 1861) very properly observes that the use of these colours ought to be interdicted, and speaking about these manufactures to his own government, he observes : " The government, whose duty it is to watch over the public health, can no longer tolerate this incessant and unwarranted sale of one of the most dangerous poisons. By the laws at present in force, a pharmacien, that is to say, an educated and responsible person, cannot sell poison without a series of formalities amounting to absolute restriction ; and should an ignorant individual be allowed to place in the hands of women and children, even more ignorant than himself, a poisonous substance, the manipulation of which leads to loss of health, and death? He who only waters the milk which he sells is condemned to fine and imprisonment, and shall a system which robs the artisan of health be allowed to continue?" In these observations and further remarks on the necessity for the legislative regulation of these manufactures, all medical men will entirely concur.

Besides these more glaring instances of clothes-poisoning, we have seen painful redness and blistering of the fingers and hands even, produced by the free contact of these parts with the common blue clothes of commerce and the cheap kid boots of the shops, and cases have been put on record of an inspector of army clothing material, and a fitter of boots, in the former of whom the finger tips were affected, and in the latter the inner side of the right hand especially was affected, by the irritant action of the dyes in the goods which they respectively handled.

Domestic remedies and their misuse.—There are certain skin troubles which result from the unadvised employment of remedies, to which attention may be conveniently directed here. We refer particularly to the misuse of liniments, plasters, and other applications.

Arnica Poisoning.—No remedy perhaps has attained a greater notoriety amongst the public than arnica. It is *the* popular remedy for strains, bruises, and similar accidents, and is universally employed by the public against them. But arnica requires to be handled with considerable care. The tincture is the preparation in general use, and the rule may be laid down that it should not only never be used in its undiluted form, but never except in a distinctly diluted form. Medical men of course will use their judgment in reference to particular conditions and cases, but the public, acting upon their own responsibility, and in a domestic fashion, should be careful not to apply any arnica tincture without dilution with water or oil. We have seen most sad and serious results follow the use of a strong preparation of arnica. In one case a gentleman had sustained a severe bruise of the outer and upper part of the thigh from a fall when playing at lawn tennis. Arnica was at once applied, and with the result of causing the leg to swell enormously from the top of the hip to the ankle, the skin becoming hot, red, glistening, irritable, and subsequently covered with innumerable blisters. There was considerable constitutional disturbance at the outset, and the patient became after a few days really ill. Abscesses threatened to form in the leg, but happily did not actually do so, but indurations of considerable size, with innumerable boils, succeeded, and the case did not really terminate in the restoration of health for at least eighteen months, so severe and disastrous was the action of the irritant arnica. We might quote many cases of a similar nature, but less severe in degree, which had come under our care at different times for treatment. It cannot therefore be too strongly urged upon the reader that the use of arnica in any form, except in a freely diluted shape, may possibly be attended with serious skin trouble. If this should occur, medical advice should at once be sought. Meantime, the inflamed part may be wrapped up in oiled rags, or freely dusted over with flour, and kept perfectly quiet.

Croton oil liniment is another application we have seen do more harm than good by misuse. Care should be exercised in the employment also of *mustard and other stimulating plasters*, especially in the young and delicate persons, for sometimes troublesome and irritable eruptions are excited by them. Lastly, we will add a word of caution concerning a remedy for ringworm that is becoming somewhat popular, viz., *Goa Powder*. If too freely applied to the head it sets up a great deal of irritation and a state simulating an attack of erysipelas of the head and face, besides a very unpleasant mahogany staining. Great care should always be taken after the use of any application to the skin to wash the hands thoroughly.

Cosmetics.—Some of these are comparatively harmless, but certainly all are not so, and probably most of them are to some extent injurious to the surface to which they are applied. The *white* substances used contain magnesia, starch, bismuth, lead, zinc, white precipitate, &c. ; the *red* are simply made up with rouge and carmine. Now most of these, if used for any time, alter the texture of the skin, and so render it hard, harsh, and coarse ; the bismuth and lead and mercury compounds give it a sallow and unhealthy aspect. None are admissible as harmless cosmetics except perhaps certain of the zinc preparations and some form of French chalk, which may be coloured by a little simple colouring matter. The late Mr. Startin was in the habit of prescribing a certain skin powder of two or three different hues, to hide blemished spots caused by disease; but these contained mercury and bismuth, and are therefore not to be recommended for use to healthy surfaces. Some of the compounds sold under the name of "milk of roses," "bloom of beauty," and the like, contain lead or bismuth. Poisoning has really occurred by the use of certain cosmetics. In the *Medical Times* for August 28, 1877, Dr. Johnson describes the case of a ballet dancer who was poisoned by the use of "flake white" as a cosmetic—this flake white containing carbonate of lead; and Dr. Rosenthal published a series of cases in a Vienna medical paper in 1876, in which he affirmed that most serious affections of the nervous system (some fatal) had resulted from the prolonged use of cosmetics of lead and mercury.

CHAPTER VI.

THE HAIR.

AMONGST the many matters connected with the cultivation of personal appearance, there is nothing that absorbs so much time and attention as the hair, especially amongst women. Yet it is strange how complete is the ignorance prevalent as to its physiology, the means of keeping it in a healthy state, and the prevention of its disorders. For as in past times some surgical operations were held to be degrading to the physician, and were consequently relegated to the barbers, "who wielded the lancet and razor alternately," so in these days most physicians have considered the hair beneath their notice, and the modern hairdresser is necessarily looked to by the public for advice and treatment when anything goes wrong with it. The reason for this state of things is that the physician busies himself more with the graver issues of life and death, and with disorders in comparison with which such affections as the loss or premature grey-ness of the hair is trifling. But although it is true that the hair is not one of the more important organs of the body, yet it is often the tell-tale of serious general mis-chief. The physician also is chary about dealing with subjects almost monopolised by a non-scientific class outside the profession and of approaching the border land of quackery, and so losing caste with his brethren by appearing to address himself to the public with a specially interested object.

If the physiology of the hair were in the least under-stood by those who have most to do with it—the hair-dressers—and they were to act on correct principles, the public would not have inflicted on it all sorts of absurd and injurious fashions, and would not be deluged, as they are, with nostrums, many of which contain elements that are useless, or perhaps, *in the long run*, productive of worse evils than those sought to be removed. If we take the most ordinary instance of hair disorder, such as its thinning and loss, this fact is abundantly exemplified. The loss of the hair is, as a rule, the consequence of debility in some form or another, but the causes of this debility are as various as the cases themselves. The usual practice is to rouse the scalp to increased action by the use of stimulant and "nourishing" applications. No doubt a certain success is sometimes obtained, but it is like spurring a jaded horse to do extra work; he may do it, but he often breaks down, and frequently knocks up after this special effort is over. So the debilitated scalp may be roused to temporary activity, but often suffers subsequently. It is the nervous system that wants strengthening by judiciously selected tonics, or the digestion that requires setting in order, and then, when there is an increased stock of energy in the system, a stimulating wash may be appropriately used. Then again there is an exaggerated notion promulgated, that the growth of the hair can be satisfactorily promoted by the rubbing into the scalp of "nourishing" applications. The reader will see in the next section how erroneous such views are; and there are many others equally so, which we shall consider in due course.

In the great majority of cases, however skilful the hairdresser may be with his scissors, it is impossible that his advice can be of great value. For as the hair is part and parcel of the body, and derives its nourishment and incentives to growth from the central organs, so it is inseparably bound up in all its states with the general health. There are few unhealthy states of the hair dependent solely on a local origin, but the condition of the hair is, as a rule, an index to the state of the body, and this of course can only be unravelled by the skilful physician.

The Structure and Functions of the Hair.—The hair, besides serving an ornamental purpose, also acts as a guard for the important underlying parts against external hurt, and as a protection against too rapid heating by the sun's rays or excessive cold.

Hairs are modifications of growth of the scarf-skin or cuticle, and grow in an oblique direction from the bottom of little depressions or tubes in the skin called *follicles*, formed in the skin as described in the preceding chapter, see figure opposite. The substance of the hairs is also composed of cells and fibres, the latter being formed in the process of growth by the flattening out of the former. The popular notion is that each hair is a tube, but this is altogether a mistake. Human hairs are always solid, the central parts, however, being made 'up of cells loosely

packed together to form what is
known as the *pith*, whilst the
surrounding outside portion is
composed of fibres or flattened
elongated cells which are densely
arranged, and overlap one another,
like the tiles on a house, to form
the cortex. Hairs are said to
consist of a *shaft*, which is the
part projecting beyond the skin ;
and of a *root*, which lies imbedded
below the surface. The root
terminates in a soft pulpy knob,
like the underground portion of
a crocus in shape, or the *bulb*,
which is situated upon one of
the little elevations or papillæ of
the true skin. This papilla is
copiously supplied with nerves
and blood-vessels bringing nutriment, and it is from this part that
the hair-cells are manufactured.
As fast as the cells are formed
they are pushed forward, as it
were, by successive additions from
below, and subsequently disposed

Hair Follicle, &c.

in layers so as to assume the form of the hair. Gradually the formed hair reaches the mouth of the follicle and protrudes above the surface of the skin. It is of great importance to understand and bear in mind that the seat of growth of the hair is deep down in the skin, and that the hair itself contains no blood-vessels and nerves, so that it does not *bleed* when cut. Near the mouth of each hair-tube or follicle two little oil glands open, which serve to keep the hair itself supple, and also the mouth of the tube and general surface of the scalp lubricated, and so prevent the too rapid passage of fluids through it. Attached also to the side of the hair are little muscles, which usually come into play, "making the hair to stand on end," only during excessive nerve stimulation, as under the influence of great cold or extreme fear. Lastly, *the colour of the hair* is due principally to the greater or less accumulation of little granules of pigment of different shades in the cells. This colouring is effected for the most part by compounds of oil with sulphur and iron; and the darker the hair the greater is the amount of iron and sulphur present.

The hair, like the rest of the body, varies in the activity of its growth at different periods of life. We know that at puberty there is an increased activity shown in very many parts of the body, and the hair participates in this reinvigoration, as seen by the growth of hair on the face. For some time the growth is continued with vigour, but after thirty or forty years of age the energy begins to flag, and in old age the formation of hair quite ceases. In extremely exceptional cases very old men have been known to cut fresh teeth and get a new crop of hair. These natural series of changes vary as to the exact time of their occurrence in different people, and are influenced by the various peculiarities and phases of constitution handed down from father to son, and by the conditions of life of the individual. That they are inevitable is the point we all are so reluctant to admit and act upon. Medical aid, however, can step in and delay progress in very many cases.

The hair, we see, varies naturally in colour, quality and quantity. Race and climate exert a well-known influence in this respect. Family peculiarities tending to produce a "wealth of hair," or a marked deficiency, or early grey-ness or loss, may be traced through the succeeding gene-rations of a family, just as features or mental conditions may be, and due allowance must be made on those points in dealing with hair disorders. So also it is a fact familiar to physicians that certain kinds of constitutions, either hereditary or acquired, are associated frequently with peculiar conditions of hair. Without seeking to maintain old views about the constant conjunction of a certain coloured hair with a special "temperament," we may instance the hair in tubercular and rickety children. The colour, quantity, and quality bear a close relationship to one another. Thus blonde hair is fine and thin in the shaft, and very thickly studded on the head; whilst black, and still more so red, hair is coarse, big in the shaft, and less thickly planted. *Curliness* of hair is mainly dependent on the degree of flattening of the shaft, for human hairs are not perfectly cylindrical. The flatter the hair the greater is the tendency to curl, and this condition is well known to be greatly affected by what is called the hygroscopic property of hair, i.e., the readiness with which it absorbs moisture from the air.

It should be remembered that it is quite consistent with health for a *very slight continuous shedding* of the hair to occur. We have already remarked that the cells of the cuticle are being continually shed, and we see natural processes in the shedding of the horse's coat, the moulting of the bird, and the casting of the snake's skin. Although then in the human being new hairs are continually springing up and old ones falling off, the process is so very slight as usually to pass unnoticed. At certain seasons, however, such as the spring and autumn, these changes go on more actively, yet then not to a great extent. When the loss at these times is excessive, then it is unnatural and the evidence of disordered health.

CHAPTER VII.

THE ORDINARY MANAGEMENT OF THE HAIR.

As in the case of the skin of the body generally, so with its more hairy parts, it is absolutely necessary that the most scrupulous *cleanliness* should be exercised. The rules to be observed in washing the body may be applied to cleansing the hair, and the reasons for these ablutionary measures apply with force, since the hair affords a special shelter for collections of "dirt" and accumulations of secretions. Most people exhibit a disinclination to wash the head, partly from fear of catching cold by the chilling of the surface through rapid evaporation of the moisture, and partly because the natural oil being washed away in the process, the hair becomes rapidly dry, harsh, and intractable. These objections are, however, easily remedied by observing to thoroughly dry and stimulate the scalp with rubbing and brushing, and then applying an artificial oil. Moreover, the scalp should have a special washing or shampooing once a week or so. Simple water or albuminous fluid may be employed, or if soap is used it must be of the mildest character, i.e., contain the least quantity of alkali possible. Brushing is one of the most valuable adjuncts to cleanliness, as it acts as an effective local stimulant, but it should never exceed reasonable limits, and irritate the scalp. As it is usual to apply some wash or pomade to dress the hair, we cannot quite pass over the subject of their use, but our remarks must be necessarily brief. These applications are of all imaginable kinds, and profess to effect every desirable object; they are greasy, or simply stimulating, or "nourishing," &c. Now, after certain principles have been laid down for his guidance, it may be left to the particular fancy of the individual to use what he likes, whether pomade or wash, scented or unscented, &c., &c. Firstly, we have seen that it is natural for the hair and scalp to be somewhat greasy, and custom has prescribed that the hair should be kept neatly arranged. A very little oil makes the hair supple, and enables it to be kept neat, and it also imparts a pleasant gloss. If any application is desired, this is all that is required in the ordinary vigorous person, and such simple applications can be easily prepared in an agreeable form. Secondly, the special stimulating elements existing in so many preparations are quite unnecessary, and indeed undesirable, except in special cases, and then they should only be used under advice. Thirdly, we have seen how and whence the hair is really nourished, viz., from the blood, and at what part it grows, viz., deep down below the surface; so that we must take for what they are worth the nourishing, renovating, or invigorating qualities claimed for so many local applications. Lastly, in whatever form the greasy or oily substance is used, it is necessary to be careful that it is quite fresh, and free from any taint of rancidity.

The hair should be cut periodically and regularly, because it is calculated that in a healthy person it grows at the rate of six or seven inches a year ; and with long hair it is so much more difficult to keep the head clean, as is well exemplified in several continental countries. The natural purposes of the hair are sufficiently fulfilled by keeping it cut reasonably short, whilst at the same time it tends to promote a more extended and vigorous growth. The practice of allowing delicate children to support very luxuriant heads of hair, however agreeable the sight may be to a fond mother, is to be deprecated, as taxing the scalp to an unusual extent at a trying period of life at the expense of the system generally. *Singeing* the hair, after it has been cut, is a process that is often strongly recommended by many empirics, but it is founded on an entirely erroneous notion, viz., that it is necessary to seal up the ends of the hair, and so prevent nourishment from draining away from the supposed tube. We have seen that the human hair is solid throughout, and does not contain any vessels, nor does it " bleed," in the sense that a plant does when cut.

There are also several *injurious customs* in vogue which may be pointed out with advantage in conclusion. In

F

women especially, care should be taken to avoid all
pressure by the wearing of tight bands across the scalp,
which partially cut off or disarrange the natural blood
supply, and so the nourishment of the hairs. In
men this is sometimes brought about by ill-fitting hats,
which compress the veins returning the blood from the
scalp, and so cause congestion of the parts. Another
injurious practice is the arrangement of the hair so that
its weight is constantly dragging on the roots of the hairs.
And again, the piling up of a mass of hair, generally
artificial, should be avoided, for it keeps the head in a
hot unhealthy condition, and leads very commonly to
localised baldness, as we shall point out later on. Many
of the hats and caps worn are unventilated, and neither
permit the proper regulation of the temperature of the
scalp, nor allow the escape of the contaminated air that
accumulates. For a similar reason the constant habit of
wearing night-caps, smoking-caps, &c., by those who have
abundance of hair, and do not want artificial protection,
is to be discountenanced.

COMMON DISORDERS OF THE HAIR.

Changes in the Colour of the Hair.—These changes
may arise from very different causes, and one or more
influences may act concurrently. Of all colour changes

premature greyness, or often rather what the individual considers so, is by far the commonest and most important. We have already considered how the hair acquires its characteristic colour from the presence of a greater or less quantity of pigment in its cells. If from any cause the supply or formation of this pigment at the bulb be interfered with or stopped the hair gradually becomes grey. Such an absence of pigment may be *congenital*, and then the hair may lose its colour in patches and other odd ways, or generally over the body in conjunction with other pigmentary changes, as in albinoes. Or there may be an *hereditary tendency*, handed down from father to son, for the supply of pigment to be cut off comparatively early in life. Or, more commonly, the lack of proper colouring matter may be due to the *state of the general health ;* for it is evident that the energy of the pigment manufactory in the hair-bulb must suffer with all other parts of the body in the decadence of the nutritive processes, or the malnutrition of ill-health, as seen in old age, after severe illnesses, under depressing nervous influences, such as fear, worry, anxiety, or hard mental work with a sedentary life, &c. And lastly, there may be a more or less local origin for the mischief, such as a persistent affection of the nerve presiding over the parts, e.g., neuralgia ; or an injury to the nerve ; or inflammation of the parts interfering with the nutritive processes; or a cutting-off of the local blood supply from disease of the blood-vessel or other cause.

With the congenital conditions and the natural changes incident to old age we have nothing to say, but the rectification of the other changes manifestly depends on a careful searching out of the exact causes of failure in the pigment supply, and a skilful general and local treatment adapted to each particular state. And moreover it is mostly these more or less remediable conditions that cause the anxious sufferer to hide his real or supposed blemish with the aid of dyes, which do not cure, but only cloak the mischief, and indeed eventually only make matters worse.

We may here take the opportunity of adding a few general remarks on the subject of HAIR DYES, RESTORATIVES, RESUSCITATORS, and the rest of the tribe of suchlike preparations. Notwithstanding quibbles and statements to the effect that they are perfectly safe and innocent, and are not dyes, the great majority of them contain poisonous substances; and it is important that the public, when about to purchase and use them, should thoroughly understand what it is about. We here largely quote the *Lancet* for January 13, 1877, which forcibly calls attention to this matter, and points out that, whereas the Pharmacy Act places arsenic and a certain number of other common poisons under legislative control, and prohibits their sale retail unless properly described, and then only under certain conditions, other poisons may be, and are, openly sold for domestic purposes without any printed warning of the risks attendant upon their misuse, or perhaps with a description leading to the inference that no poison is present. The *Lancet* purchased and submitted to the independent analysis of two well-known chemists twenty-one samples of the preparations most extensively advertised and in use in England and America. These gentlemen found that of eighteen preparations *for darkening the hair*, sixteen consisted (omitting comparatively unimportant ingredients, such as glycerine and scent) of su'phur and a considerable quantity of *lead*, the seventeenth had lead without the sulphur, and the eighteenth effected its purpose by a combination of ammonio-nitrate of silver and pyrogallic acid. The remaining three preparations were intended for *lightening the colour* of the hair. They were found to be infinitely less dangerous to the general health at any rate, and were substantially identical, containing a tolerably concentrated and slightly acidulated solution of peroxide of hydrogen. The German Sanitary Board act vigorously in these matters, and publish the results of the analyses for the use of the public. In No. 38 of their *Veröffentlichungen* the analyses of two well-known preparations is given, and the warning appended, to the effect that they are "highly dangerous to health." There are, in addition, other very numerous and well-known methods for darkening and bleaching and giving a golden hue to the hair, but as from various reasons they are little used now we shall not stop to discuss them further than to say that the black dyes mostly contain salts of silver, mercury, and iron.

Very conflicting statements have been made from time to time with regard to the injurious effects produced by the absorption of these dyes. Whilst some have denied the possibility of the system getting affected, others have described almost sensational effects. We think that it cannot now be denied that deplorable results do occasionally ensue, and they are none the less serious because often so insidious; in any case their action on the nutrition of the hair is certainly bad. Such being the case, if the public are determined to use dyes to cloak their blemishes, and approximate themselves to standards of natural or ideal beauty, which they have set up, then the Legislature, whose duty it is to concern itself with the PUBLIC HEALTH, ought to see that complete information as to the nature of the preparations is appended, and knowledge of their effects attainable.

The evil effects of the free use of these lead dyes is often not noticed because not immediately seen. The *Lancet* remarks that—"*Many recorded cases show that very minute quantities of lead may after a time produce symptoms of poisoning.* Certain circumstances, moreover, induce us to think that incipient lead-poisoning is more common than is generally supposed. In all chemical laboratories the testing for lead in drinking-water is a common experience. The number of samples of water sent for this purpose is surprising. Now, in a great many instances no lead is found, and it is worthy of consideration whether in some of these cases the symptoms which threw suspicion unjustly on the water may not have been caused by the use of lead cosmetics." And Dr. Garrod, F.R.S., a physician eminently qualified to speak on this topic, says, whilst writing on lead-poisoning (*Lancet*, vol. i. 1872)—"As to the power of the skin, when un-injured, to absorb lead, some doubts have existed. It was altogether denied by Tanquerel des Planches ; but, nevertheless, there are many facts which seem to favour the idea that it may be thus introduced. The recent common use of hair dyes, which very constantly contain a large amount of lead, has appeared to cause in many patients effects which probably are dependent on the absorption of this metal ; and although in such cases the action at any one time is exceedingly slight, still the use of such dyes is generally continued for a length of time, and may thus slowly lead to injurious consequences. I have certainly met with patients suffering from headaches and other uneasy feelings who have ceased to experience these symptoms when they have given up the applications of the dyes to their hair." It should further be remembered that gouty people, whilst probably frequent users of dyes, are peculiarly susceptible to the action of lead. Another very distinguished medical man stated to us his strong conviction that in many old dandies the free use of lead hair-dye had been followed by odd semi-paralytic symptoms in connection with the bladder and other parts. The use of these lead dyes, then, if determined on, should be exercised with the greatest caution, and in the weakest solutions. Many people think that when once applied the hair is dyed for ever, but one great evil is, that when once used, the application of the dye must be continuous, for that part of the hair only is dyed which is above ground, so to speak, and the new portion of the shaft of the hair pushing its way above the surface appears of the original colour. There can be no great objection, however, to the use of a little mild sulphur pomade or wash, if the gradual darkening of the hair is thought desirable. That the hair finally suffers in the vigour of its growth from the prolonged use of hair dyes is unquestionable in our experience.

Superfluous Hair.—There may be an unusually luxuriant growth in natural situations, such as the hair of a woman's head, the moustache and beard of a man, and

on the male body generally. Many remarkable examples
are on record of such conditions; but what we have
specially to deal with here is the growth of hair in
unusual situations and in an anomalous manner, due
either to congenital causes, eccentricities of nature, or
disease. For instance, *hairy warts* and *moles* are common
enough on the skin as congenital conditions, and they
may get much larger in after life. Hairy patches may
spring up after irritation of a part, as after a blister ; and
hair may grow in anomalous luxuriance on the female
body, as in a strange case related of a young lady, who
acquired a short hairy coat during convalescence from a
severe illness. Lastly, we may mention the growth of a
beard or moustaches in women, occurring often in
connexion with ill health.

In the treatment of these conditions for their removal,
the greatest nicety and care is required. The amount
of relief which can be afforded depends on the nature
and extent of the hairy growth, whether a wart or mole
or of another description. It must be borne in mind
that the hair manufactory is deep down in the true skin,
and the problem is how to destroy the former or stop its
action without injury to the surrounding skin, or production of a sore and scar, i.e., with the least deformation
possible. It is easy to destroy the shafts of the hairs and
the cuticle, and so much without scarring, but, as the

papillæ producing the hairs are still left, the latter only grow up again ; and so it is when the hairs " are pulled out by the roots." The preparations commonly used are called " depilatories," and they are numerous, and some of ancient date. They are not to be used without due appreciation of their mode of action and directions for their use, and also with the knowledge of the degree of service they can be in ordinary cases. In complicated cases other operations of great nicety are sometimes possible, but where very extensive tracts of hair exist the removal is almost impossible.

Thinning of the Hair and Baldness.—We might easily compose a fairly long treatise upon this topic of so much interest to a great number of people, but space will only permit a general summary of the causes and means of cure. It has been said that "to remedy baldness and thinning of the hair it seems by the daily advertisements that every advertiser had ransacked the whole arcana of science and discovered a secret process ;" but the specifics so much vaunted are in most cases doomed to disappoint, for the simple reason that the causes of the disorder are to be found in the state of the general health, which these local panaceas fail to influence. It cannot be too strongly or frequently insisted on that loss of hair is the result but rarely of purely local causes, and that consequently all attempts to remedy it by local measures alone are not likely to be attended with much success; nay, they often do positive harm.

Just as in advancing life the supply of pigment or colouring matter of the hair gradually falls short and finally ceases, so in like manner the quantity of hair diminishes, i.e., the cell manufactories lose their energy, until over the greater part of the scalp the hair is shed not to be reproduced, and the hair-forming apparatus itself atrophies and shrivels away. During the whole period of life there is the continual slight shedding going on of the old hairs and replacement by new ones, and, in a healthy state and the vigour of life, the balance between the two is preserved; but if from any cause, e.g., debility, the formation of new hair is checked, then there results, of necessity, thinning or baldness. The age at which the energy of the growth of hair declines as a natural process, varies in different individuals, and hereditary peculiarities exercise a very marked influence. It is quite natural for the members of certain families to become bald early in life. There is difference between the natural baldness or thinning of the hair, and that which results prematurely from disease; for the former is inevitable, and cannot be influenced or stayed to any great extent by remedies, whereas the latter is mostly preventable and remediable by medical aid. The problem, then, which we invariably set ourselves to solve in reference to these cases is whether the loss of hair is in any degree a natural occurrence; because, where it is so, the promise of restoration of the growth to its previous condition must necessarily be greatly qualified. The promise of full restoration would be an imposition, and only makes the pocket suffer at the hands of ignorant empirics. Something may, it is true, be done by the judicious use of stimulating applications, but care must be taken that they are not used in too great strength. Happily the bulk, however, of cases of thinning of the hair and baldness are capable of improvement if due attention be given to the general health conditions.

There are three main groups of causes which we will first give a general glance at, and then treat more in detail.

1. *Local causes*, which interfere with the nutrition of the hair, such as inflammation of the skin and oil glands, e.g., Eczema and Seborrhœa; or the attack of vegetable parasites, as in Ringworm; or a local cutting off of the blood supply.

2. *General causes*, which lead to debility of constitution, and so indirectly weaken the nutrition of the hair. In this group may be included all severe diseases that weaken the bodily powers, e.g., during convalescence from fevers; inflammation of important internal organs, as the lungs; after childbirth; during the progress of consumption; in

special forms of blood-poisoning ; in prolonged derange-
ment of the digestive organs ; in connexion with debility
from continued excesses of various kinds ; in the weakness consequent upon over-anxiety, prolonged worry,
hard work with sedentary life ; residence in relaxing
climates, and so forth.

3. *Purely nervous causes.*—Although it is true that
under ordinary circumstances, where debility is present,
there is a lack of proper nerve influence over the processes of nutrition of the hair, yet in some cases the want
of this stimulation, or, to use a popular expression, nerve
electricity, seems to be the direct and only cause of the
thinning or baldness.   The special implication of the
nerves is evidenced in various ways, such as neuralgic
pains of the scalp, and the loss of hair may be only local,
or even general and absolute.

Such are the main causes of loss of hair ; and it must
be at once evident that, in attempting to remedy the disorder, no one plan of treatment can cope with such mul-
titudinous causes. No mere local panacea or "stimulating
wash," or "hair-restorer" rubbed into the scalp can effect
the desired result, but the treatment must consist of something beyond mere local measures, and must vary with the
exact cause of the mischief.   Acute and keen judgment
must be brought to bear upon the matter to unravel the
exact one out of a variety of influences.

To go a little more into detail, firstly as regards local causes, the treatment is clear and definite. Persons usually consult the medical man for a particular eruption or ringworm, which leads to the loss of hair, and the latter begins to be remedied so soon as the eruption itself is cured. There is a prevalent fashion, however, which, as leading to loss of hair, should be noticed here. The reference is to the practice of wearing a mass of hair, often including a quantity of pads and ornamental adjuncts, tightly at the top or back of the head. The pressure exerted by the latter, and the heat consequent upon their use, lead, in many cases, to inflammation of the roots of the hairs, and subsequent baldness. The remedy is to give the hair and scalp fair play for a time; to use soothing remedies in the first instance, and avoid all stimulating washes, as calculated to increase the mischief; and when the scalp gets into a more healthy state, and not till then, to prescribe some mild application to make the hair grow again.

Dandriff, or Scurfiness, with general thinning of the hair, is one of the most common conditions of the scalp for which advice is sought. The term *dandriff* is the popular expression for the existence of a dry, scaly condition of the scalp, which is very common, and causes great annoyance by covering the shoulders with the falling of the branny scales, especially when the head is brushed, and by itching and heat. It is associated also with thinning of the hair in long-standing cases, and is one of the affections for which the public rush to the hairdresser or "hair doctor" for advice and a nostrum. The term includes two distinct diseases of the scalp, which should be treated according to their nature and causation. *Firstly*, there is the superabundant formation and shedding of cuticle induced by an excessive blood supply to the parts, set up by too hard brushing, the use of washes of too stimulating a character, by rancid pomades, the neglect of cleanliness, &c. It is also caused by loss of tone in the vessels due to constant indigestion, excessive heating of the scalp by hats, and caps, and hair, and by keeping the scalp too dry or devoid of oil. *Secondly*, there is the presence of little plates or scales of congealed fatty matter formed in excessive quantity and altered quality by the deranged or inflamed oil glands. In this affection the fatty scales as they form must be prevented from accumulating by suitable softening and solvent applications, and the formation stopped by astringent remedies and the bringing back of the oil glands to a properly healthy state.

In very many cases the disease attacks young ladies who are weak and more or less delicately constitutioned; under such circumstances it must be dealt with by cod-liver oil and iron internally, and mild astringents externally. In other cases, middle-aged ladies suffer; and in these instances the disease is often associated with an acid state of system of a gouty tendency, which needs special anti-acid remedies internally. A borax and camphor wash is a useful one to try, but if not successful proper medical advice should be sought.

Thinning caused by ringworm will be dealt with in a separate section.

As regards the *general* causes of thinning and baldness, each case must be dealt with on its merits. The particular cause of debility must be tracked, recognised, and dealt with accordingly. The use of stimulants to the head must be carefully timed, and not used until the disorder of the system is in some measure recruited. If used too early, or without proper restrictions, they only do harm. We cannot, of course, give any set of remedies here for such diverse conditions, but we may mention that a very satisfactory wash and restorative under most conditions of thinning and loss of hair is composed of the tinctures of nux vomica and lytta, distilled vinegar, with glycerine, honey, and rosewater in due proportion.

Lastly, as regards the disorder, when traced to a special and marked nervous causation, it may be affirmed that amongst ladies few who have much anxiety and trouble escape a certain amount of thinning of the hair as the *result of nervous exhaustion*, but this is remediable

by the judicious use of nerve tonics. There is one particular form of baldness due to a failure of the proper
nervous influence over the nutrition of the hair and scalp,
which must not be overlooked in this place. It occurs
in circular patches on the scalp, and is popularly regarded
as a species of ringworm. It is common in the young,
especially in growing school-girls, but also boys, and it
may be seen later on in life. Therefore it occurs mostly
at a time when the powers are strained during the rapid
growth and development of the body, and this outlying
part suffers accordingly. The first thing that attracts
attention is a small bald spot, from whence the hair has
suddenly fallen. This spot increases week by week, and
meantime one or several other places begin to manifest
their presence; so that in marked instances of the malady
the scalp is studded over with small bald spots, varying
in size from the area of a shilling or half-crown to a small
hand palm, when the spots coalesce. The bald patches
look quite white and polished, like a billiard ball, and
are deprived more or less of their natural sensibility.
The hair which falls out has atrophied and tapering roots,
and shows evidence, to the skilled eye, of faulty nutrition.
In some extreme cases the scalp is almost devoid of
hair, or indeed completely bald, the eyebrows and other
hairy regions participating. We have seen cases where
the loss did not take place as it usually does, gradually,

but rapidly in a few weeks. The causation is found in these cases in the "overgrown" child, in the system overtaxed by excesses or debilitated by mental anxiety, prolonged and assiduous study for competitive examinations and other depressing causes. In the treatment, the exact cause or phase of nerve trouble must, as far as possible, be made out and dealt with by internal remedies. Occasionally severe neuralgic symptoms point to quinine, or minute doses of arsenic and suchlike drugs, as eminently suitable. Locally, in connexion with the internal tonics, these cases are to be dealt with by stimulants, judiciously regulated, to rouse the topical nerves, by suitable frictions and the continuous electrical current. If the scalp has lost its sensitiveness to a marked extent, at a certain stage blistering may be advisable, and should be followed by the use of strong stimulants well rubbed in night and morning. A warning may be given that a great deal of patience will have to be exercised by the sufferer, since the progress of the disease towards cure is always slow. But if the patient can be prevailed upon to place confidence in the continued use of properly-selected remedies, instead of embracing nostrums from every quarter, and if he be in fair health and respond to tonics, a cure may be always promised. If, however, the patient be of an advanced age, and the scalp be shining and glistening, and extensively devoid of hair, such a favourable result is very doubtful. But even in apparently hopeless cases of baldness, if they be not of too long standing, the return of growth of the hair to its natural or almost natural state, may be witnessed with due perseverance and carefully-planned treatment.

Ringworm.—Of all the remediable causes of thinning of the hair and baldness ringworm is the most important, both on account of the extreme frequency of its occurrence, and the annoyances and vexations attendant upon it. Unlike the other causes of diseased or lost hair, it is catching from one person to another, so that the child affected with it has to be in a measure shut out from the free intercourse with his playmates, and placed under special restrictions in association with the family and neighbours. Differing in another way also from some of the other common affections, the common ringworm of this country is no respecter of persons, and occurs with great frequency amongst the well-to-do classes. Occurring, too, as it does, in by far the majority of cases, at an age when the training of the mind is of the first importance, it operates in a peculiarly unfavourable and vexatious manner in keeping the child away from school. There is indeed no disease to which we are subject that, being in itself of a comparatively trifling nature as regards its influence on the health, causes so great anxiety to parents, and of which so much dread is entertained.

There are several varieties of ringworm, but they are caused by some member of those lowest forms of vegetable life called *fungi*. We shall here describe particularly the variety that is almost invariably met with in England, and only allude incidentally to the other kinds which are very rare here. The disease is caused by the little spores of the fungus alighting and settling on the hairy scalp where they immediately commence to multiply and grow, insinuating themselves and penetrating between the cells of the cuticle and fibres of the hair. This causes the fibres to separate and the shaft to look dry, dull, opaque, swollen, and brittle, so that it readily breaks off. From the shaft the fungus spreads downwards between the fibres to the root and there firmly establishes itself; and, as it interferes with the proper nutrition of the hair, and steals and grows on what nutriment there is, the hair gradually withers up and becomes distorted, or dies and is shed. If there is life enough for new air cells to be formed these are rapidly affected. From the original spot of infection the fungus spreads, just like the " fairy circles " on grass, infecting hair after hair, until *usually* a more or less circular patch is formed (hence the name "ringworm"), which, if not actually bald, has the hairs so altered in appearance as immediately to attract attention. Out of this mode of growth the error has sprung up amongst the public that every circular patch of eruption must be ringworm, and also that any red scurfy patch that is not circular is therefore not this disease.

The first thing generally noticed is *itching* of the head, which is complained of by the child, or the frequent scratching is noticed by the mother or nurse. Sometimes this symptom is not prominent, or goes unnoticed, and a little scurfy altered patch is suddenly and unexpectedly come upon in arranging the hair. It is quite remarkable what a length of time a staring patch of ringworm, perhaps as big as a half-crown piece, will remain unnoticed, even by suspicious and anxious people in good circumstances. Whatever leads to its detection, when examined, it is seen to be of a peculiar dull hue, ragged-looking, more or less circular, and to differ altogether in aspect from the surrounding healthy parts. The individual hairs on the patch are short, swollen at their bases, and, because so brittle, they get their ends broken off, and look like as if something had nibbled at the ends. They are seen to have lost their usual gloss and lustre, and to be dull-looking and opaque, and also to have undergone a very peculiar corkscrew twist. These diseased hairs lie like beaten-down corn mixed up with, and partially hidden by, any healthy erect hairs that may be present, especially in the early stages. The intermediate skin, which is also attacked, becomes "coarse" and dry, and generally a lot of scales are thrown off, making the patch scurfy. Such is the character of an ordinary case of ringworm after it has been spreading a short time. If seen at a very early stage perhaps simply a few white scales on a tiny reddish spot of scalp around a dull lustreless hair would be present, which condition no one but a skilled or experienced person would understand certainly. This spot spreads equally and in all directions, usually contrasting in its dull, ragged, scurfy characters with the surrounding hair, until it reaches the stage described above. Later on the patch may become almost entirely bald ; or the irritation produced by the fungus may inflame the hair-follicle, and the hair may appear issuing from a yellow pimple full of matter ; whilst the whole patch may be hot, red, and swollen. Many distinct patches may arise on the same head from infection by scratching or otherwise, and these may join here and there until the whole scalp almost appears diseased. There is a form of ringworm called Favus which especially selects the poor for its attacks, and is much commoner in Scotland and on some parts of the Continent than in England. In Favus large crusts are formed on the head which assume a characteristic honeycomb appearance.

Like all vegetable growths, the ringworm fungus has an especial predilection for certain qualities of skin, i.e., the skin of children of a peculiar delicate organisation ; and it is well known to flourish with extraordinary luxuriance in the hot and moist tropical countries. The children particularly affected are those of delicate pallid "lymph-atic" constitution, if not actually strumous and phthisical.

As to the source from which the ringworm comes, it is impossible to track its path very often, but usually it is handed on from one head to another by pretty direct contact, or acquired from horses, pigs, dogs, cats, or mice. Though confined in its attacks almost entirely to children, it is necessary to remember that adults occa-sionally do catch it in the hands and elsewhere, so that proper precautions in attendants are necessary.

In the treatment of this affection the greatest patience is required on all hands, and complete confidence in the doctor. The results attained are often eminently unsatis-factory; for frequently the doctor treats this apparently trivial affection with contempt and repugnance, and the parent, impatient of success and deluged with nostrums from friends, flies from one to another, and fails to carry out a thorough and sure mode of treatment. The doctor dislikes having anything to do with it, because if he cures the disease he gets no credit, and if, as is often unavoid-able, it baffles him for a long time, it is a source of disgust to him and of great concern to the friends. The difficulty experienced in the cure is instanced by the vast number of nostrums and specifics which exist on every hand, from the popular ink and vinegar upwards; yet in by far the majority of cases, if the cure is set

about on proper lines, the disease can be soon era-
dicated.

To set down the directions for cure in all the variations
and phases of ringworm would occupy too much space,
so we will content ourselves with considering a few
leading principles.

Firstly, it is necessary to keep steadily in mind that to
cure the disease you have to destroy a fungus which is
constantly tending to spread and infect healthy parts,
and recover lost ground. *Secondly*, it is of the utmost
importance that the disease should be recognised early,
and energetic measures taken before the fungus has
spread down below the surface of the skin to the roots.
If taken early, while still superficial, a great number of
applications, such as ink, vinegar, tincture of the per-
chloride of iron, sulphur, iodine, carbolic acid, a blister,
goa powder, a mild mercurial ointment, sulphuric acid,
&c., &c., will readily cure the disease, and easily earn
a widely spread reputation. But once let the fungus get
down the shaft into the roots and the difficulties commence. Therefore the lesson to be learnt is, that *every*
mother in her nursery, and every head of a school or insti-
tution where young boys or girls are congregated, should
enjoin the frequent examination of the heads, to detect early
any possible outbreak of ringworm.

Thirdly, if the disease has reached a later stage and

the roots are implicated, it is of the last importance to appreciate the fact that only by *persistent* and *continuous* treatment can it be cured. The duration of the disease is to a great extent in the friends' hands. Over and over again it happens that a doctor's explicit directions are given an imperfect, because uninterested, execution by servants, and then, as no progress is made, every one gets disgusted and wearied. Hence a great stride is made when a doctor succeeds in exciting a real interest in the cure and a thorough understanding of the methods used and objects wished to be obtained. We will mention as examples *epilation* and the *application of remedies*. If the hairs which are the seat of the growth of the fungus can be pulled out, or epilated as it is called, the process is a very useful adjunct to cure. The difficulty is to do so without breaking them off, as they are so brittle, and if imperfectly carried out it is useless or worse. Therefore it is important to teach the attendants to recognise diseased hairs from healthy ones and see that they understand how to draw them out properly. Again, a great art in the application of remedies is so to adjust their strength to the sensitiveness of the scalp as to excite a *slight inflammation* of the follicles, which loosens the hairs in their sheaths, and permits their easy extraction and falling out. But, on the other hand, as there is the urgent necessity that the rubbing in remedies should be *thorough*, and carried on *without a break* to prevent the disease getting a respite and recovering its lost ground, it is also necessary on this head to adjust the strength of the remedies to prevent too severe inflammation, and a consequent enforced break in the treatment. An intelligent, well-instructed, interested attendant will manage these matters, and make all the difference in the success of the treatment.

Lastly, the error is of frequent occurrence and disastrous in its consequences, to pronounce a ringworm cured before such is really the case. None but a skilled eye can safely determine the fact. Partly through weariness of the case in all parties, and the belief that the few remnants of disease will surely die out or do no harm, the remedies are discontinued, and the boy is perhaps sent off to school, and the girl allowed to mix freely with her playmates, with the only too certain consequence of its reappearance. We see this over and over again. As long as a single diseased hair remains, or stump of a hair, the ringworm is not certainly cured.

In schools and other institutions the greatest precautions ought always to be in force of a preventive nature, and when a case of ringworm occurs complete isolation should be carried out.

Besides the remedies mentioned above, a multitude more are in use for the cure of ringworm, such as sulphurous acid, tarry oils and creosote, glacial acetic acid, croton oil, Coster's paste, &c. They are all either kept applied to the head as *lotions* or rubbed into the scalp thoroughly as ointments compounded with lard or vaseline. For further details we must refer the reader to special treatises on ringworm and skin diseases.

The few words that *Ringworm of the Body* requires from us may be added here. It may attack the body primarily or be transplanted to the hands or trunk from the head. It begins as a reddish, itching, scaly spot, which spreads and clears up in the centre, so that a ring is generally formed. Sometimes the ring has a lot of little clear bladders or vesicles around its border. The treatment is very simple. A few light applications of ink, acetic acid, sulphur, iodine or carbolic acid suffices as a rule to cure the disease.

Pediculi or Lousiness of the Head.—With the more cleanly fashions of wearing the hair now in vogue in this country, and with the prevalence of luxuries conducive to cleanliness dependent on the advance of civilisation, this affection is practically confined to the poorer classes of society. Amongst these it is of the commonest occurrence, and indeed it is hardly exaggeration to say that the majority of the poorer classes of hospital patients are affected with the disease, especially the children, who rarely get their heads washed or brushed. The children, however, of the richer classes do occasionally contract the affection at school, and from servants or poorer playmates. These objectionable parasites, called " pediculi," run about on the scalp and deposit their cup-like eggs, known as " nits," on the lowermost portion of the shaft of the hair. The disturbance they produce in a direct manner in their travels, and by their sucking at the " pores," may be comparatively insignificant, but the irritation induces scratching, and then the secondary ill-effects vary according to the state of health of the patient. If the child be badly nourished, scrofulous, or otherwise debilitated, sore places are induced, which scab over, and the discharge mats the hairs together. These sore places in their turn set up the enlargement and inflammation of the glands of the neck, forming large lumps, which may go on to abscesses. Cutting short off all the hair necessarily removes the majority of the parasites and their eggs, and their harbour. This is only necessary, however, in the severe cases. The head should be thoroughly greased with a weak white precipitate or stavesacre ointment, containing some strong scent, and the application allowed to remain on all night. In the morning a good washing with soap and water, and a re-application of the ointment is to be made. A few applications will effectually destroy all parasites, but not the eggs. These must have their attachments softened by a thorough wetting for half an hour with weak vinegar and water (1 to 4), and then be well washed off.

Scabbed Heads.—Scabs or crusts may be formed on the head in the course of several distinct diseases, which are from the first, or become, of an inflammatory nature. The thick, yellow, greasy plates formed in the disease called *seborrhœa* may be mistaken for crusts, and to this affection we have already drawn special attention. True scabbing is caused by the drying up of more or less mattery discharges or " weepings," which become solid just as blood forms a clot. It occurs in all the forms of running *eczema* or *impetigo* of the head ; in some forms of ringworm, especially seen in the yellow honeycombed crusts of *favus ;* and as a consequence of any " sore " on the head, notably such as are produced by the scratching due to the irritation of lice. The more " matter " or pus discharged, the larger will be the dirty yellow crust ; and it is very important to understand that much more matter is discharged, and there is less tendency for the sore to heal, in delicate children and those who are ill-fed and out of health, especially the scrofulous. To get the head well the child (for children are mostly affected) must have as much good food as possible, and tonic and nourishing medicine must be given. As for the head itself, the first thing to do is to get the scabs off by poulticing or soaking in oil (the hair being cut short if rendered necessary by the quantity of scab), and then to apply a soothing and slightly astringent ointment, such as is afforded by mixing up five grains of white precipitate in an ounce of fresh lard.

LONDON : PRINTED BY WILLIAM CLOWES AND SONS, STAMFORD STREET
AND CHARING CROSS.

BOGUE'S HALF-HOUR VOLUMES.

THE GREEN LANES: A Book for a Country Stroll. By
J. E. TAYLOR, F.L.S., F.G.S. Illustrated by 300 Woodcuts. Fifth
Edition. Crown 8vo., cloth, 4s.

THE SEA-SIDE; or, Recreations with Marine Objects. By
J. E. TAYLOR, F.L.S., F.G.S. Illustrated with 150 Woodcuts. Fourth
Edition. Crown 8vo., cloth, 4s.

GEOLOGICAL STORIES: A Series of Autobiographies in
Chronological Order. By J. E. TAYLOR, F.L.S., F.G.S. Numerous Illustrations. Fourth Edition. Crown 8vo., cloth, 4s.

THE AQUARIUM: Its Inhabitants, Structure, and Management. By J. E. TAYLOR, F.L.S., F.G.S. With 238 Woodcuts. Crown 8vo.,
cloth extra, 6s.

THE MICROSCOPE: A Popular Guide to the Use of the
Instrument By E. LANKESTER, M.D., F.R.S. With 250 Illustrations.
Sixteenth Edition. Fcap. 8vo., cloth plain, 2s. 6d.; coloured, 4s.

THE TELESCOPE: A Popular Guide to its Use as a means
of Amusement and Instruction. By R. A. PROCTOR, B.A. With numerous
Illustrations on Stone and Wood. Fifth Edition. Fcap. 8vo., cloth, 2s. 6d.

THE STARS: A Plain and Easy Guide to the Constellations.
By R. A. PROCTOR, B.A. Illustrated with 12 Maps. Tenth Edition.
Demy 4to., boards, 5s.

ENGLISH ANTIQUITIES. By LLEWELLYNN JEWITT,
F.S.A. Illustrated with 300 Woodcuts. Crown 8vo., cloth extra, 5s.

ENGLISH FOLK-LORE. By the Rev. T. F. THISELTON
DYER. Crown 8vo., cloth, 5s.

PLEASANT DAYS IN PLEASANT PLACES. Notes
of Home Tours. By EDWARD WALFORD, M.A., late Scholar of Balliol
College, Oxford, Editor of "County Families," &c. Illustrated with
numerous Woodcuts. Second Edition. Crown 8vo., cloth extra, 5s.

Other Volumes in preparation.

London : DAVID BOGUE, 3, St. Martin's Place, W.C.

USEFUL AND INTERESTING WORKS.

A complete Catalogue gratis on application.

THE ANIMAL KINGDOM. By Baron Cuvier. Edited with considerable additions by W. B. Carpenter, M.D., F.R.S., and J. O. Westwood, F.L.S. New Edition, illustrated with 500 Engravings on Wood and 36 Coloured Plates. Imp. 8vo , cloth, 21s.

A PLAIN and EASY ACCOUNT of the BRITISH FUNGI. By M. C. Cooke, M.A., LL.D. With especial reference to the Esculent and other Economic Species. With Coloured Plates of 40 Species. Third Edition, revised. Crown 8vo., cloth, 6s.

RUST, SMUT, MILDEW, AND. MOULD. By M. C. Cooke, M.A., LL.D. An Introduction to the Study of Microscopic Fungi. Illustrated with 269 Coloured Figures by J. E. Sowerby. Fourth Edition, with Appendix of New Species. Crown 8vo., cloth, 6s.

THE PREPARATION and MOUNTING of MICRO-SCOPIC OBJECTS. By Thomas Davies. New Edition. Edited by J. Matthews, M.D., F.R.M.S. Fcap. 8vo., cloth, 2s. 6d.

A NEW LONDON FLORA ; or, Handbook to the Botanical Localities of the Metropolitan Districts. By E. C. De Crespigny, M.D. Compiled from the Latest Authorities and from Personal Observation. Crown 8vo., cloth, 5s.

SYNOPSIS FILICUM ; or, A Synopsis of all Known Ferns, accompanied by Figures representing the Essential Characters of each Genus. By Sir W. J. Hooker, F.R.S., and J. G. Baker, F.L.S. Second Edition. 8vo., cloth, £1 2s. 6d., plain : £1 8s., coloured.

THE COLLECTOR'S HANDY-BOOK of Algæ, Diatoms, Desmids, Fungi, Lichens, Mosses, &c. With Instructions for their Preparation, &c. By Johann Nave. Translated and Edited by Rev. W. W. Spicer, M.A. Illustrated with 114 Woodcuts. Fcap. 8vo., cloth, 2s. 6d.

BRITISH BUTTERFLIES AND MOTHS, an Illus-TRATED NATURAL HISTORY OF. With Life-size Figures of each Species, and of the more striking Varieties. By Edward Newman, F.Z.S. 1000 Woodcuts. Super-royal 8vo., cloth gilt, 25s.

FLOWERS : Their Origin, Shapes, Perfumes, and Colours. By J. E. Taylor, F.L.S., F.G.S. Illustrated with 32 Coloured Figures by Sowerby, and 161 Woodcuts. Second Edition. Crown 8vo., cloth gilt, 7s. 6d.

NOTES ON COLLECTING AND PRESERVING NATURAL HISTORY OBJECTS. Edited by J. E. Taylor, F.L.S., F.G.S. With numerous Illustrations. Crown 8vo., cloth, 3s. 6d.

London : DAVID BOGUE, 3, St. Martin's Place, W.C.

USEFUL AND INTERESTING WORKS.

A complete Catalogue gratis on application.

THE SIGHT, and How to Preserve It. By H. C. ANGELL, M.D. With numerous Illustrations. Fcap. 8vo., cloth, 1s. 6d.

BOOK OF KNOTS. Illustrated by 172 Examples, showing the manner of Making every Knot, Tie, and Splice. Third Edition. Crown 8vo., cloth, 2s. 6d.

ROYAL GUIDE TO THE LONDON CHARITIES, 1878-9. By HERBERT FRY. Showing, in alphabetical order, their Name, Date of Foundation, Address, Objects, Annual Income, Chief Officials, &c. Sixteenth Annual Edition. Crown 8vo., cloth, 1s. 6d.

THE CARE AND CURE OF THE INSANE : Being the Reports of *The Lancet* Commission on Lunatic Asylums, 1875-6-7, for Middlesex, City of London, and Surrey (republished by permission), with a Digest of the principal records extant, and a Statistical Review of the Work of each Asylum, from the date of its opening to the end of 1875. By J. MORTIMER-GRANVILLE, M.D., L.R.C.P. Two Vols., demy 8vo., cloth, 36s.

A MANUAL OF BEE-KEEPING. Containing Practical Information for Rational and Profitable Methods of Bee Management. By J. HUNTER. With Illustrations. Second Edition. Fcap. 8vo., cloth, 3s. 6d.

FOOD CHART, giving the Names, Classification, Composition, Elementary Value, rates of Digestibility, Adulterations, Tests, &c., of the Alimentary substances in general use. By R. LOCKE JOHNSON, L.R.C.P., L.R.C.I., &c. In wrapper, 4to., 2s. 6d.; or on roller, varnished, 6s.

OUR FOOD : Lectures delivered at the South Kensington Museum. By E. LANKESTER, M.D., F.R.S., F.L.S. New Edition. Illustrated. Crown 8vo., cloth, 4s.

THE USES OF ANIMALS in Relation to the Industry of Man ; Lectures delivered at the South Kensington Museum. By E. LANKESTER, M.D., F.R.S., F.L.S. New Edition. Illustrated. Crown 8vo., cloth, 4s.

THE PRINCIPAL PROFESSIONS, a Practical Hand-BOOK TO. Compiled from Authentic Sources, and based on the most recent Regulations concerning admission to the Navy, Army, and Civil Services (Home and Indian), the Legal and Medical Professions, the Professions of a Civil Engineer, Architect and Artist, and the Mercantile Marine. By C. E. PASCOE. Crown 8vo., cloth, 3s. 6d.

London : DAVID BOGUE, 3, St. Martin's Place, W.C.